The Cyanide Process of Gold Recovery

By A. Scheidel

with an introduction by Kerby Jackson

This work contains material that was originally published in 1894.

This publication was created and published for the public benefit,
utilizing public funding and is within the Public Domain.

This edition is reprinted for educational purposes
and in accordance with all applicable Federal Laws.

Introduction Copyright 2014 by Kerby Jackson

Introduction

It has been one hundred and twenty years since the State of California released it's important publication "The Cyanide Process: Its Practical Application and Economical Results". First released in 1894 this important volume has now been out of print for over a century and has been unavailable to the mining community since those days, with the exception of expensive original collector's copies and poorly produced digital editions.

It has often been said that "*gold is where you find it*", but even beginning prospectors understand that their chances for finding something of value in the earth or in the streams of the Golden West are dramatically increased by going back to those places where gold and other minerals were once mined by our forerunners. Despite this, much of the contemporary information on local mining history that is currently available is mostly a result of mere local folklore and persistent rumors of major strikes, the details and facts of which, have long been distorted. Long gone are the old timers and with them, the days of first hand knowledge of the mines of the area and how they operated. Also long gone are most of their notes, their assay reports, their mine maps and personal scrapbooks, along with most of the surveys and reports that were performed for them by private and government geologists. Even published books such as this one are often retired to the local landfill or backyard burn pile by the descendents of those old timers and disappear at an alarming rate. Despite the fact that we live in the so-called "Information Age" where information is supposedly only the push of a button on a keyboard away, true insight into mining properties remains illusive and hard to come by, even to those of us who seek out this sort of information as if our lives depend upon it. Without this type of information readily available to the average independent miner, there is little hope that our metal mining industry will ever recover.

This important volume and others like it, are being presented in their entirety again, in the hope that the average prospector will no longer stumble through the overgrown hills and the tailing strewn creeks without being well informed enough to have a chance to succeed at his ventures.

Kerby Jackson
Josephine County, Oregon
November 2014

PREFACE TO FIRST AND SECOND EDITIONS.

Hon. J. J. Crawford, State Mineralogist, San Francisco, Calif.:

Dear Sir: In accordance with your letter of December 8, 1893, I herewith submit my report on the cyanide process. I have endeavored to describe that process in its practical application and economical results. The information it conveys includes my own experience, and is supplemented from articles which appeared in technical periodicals; also from the records of patents granted by the Patent Offices of the United States and Great Britain, and from the Blue Books issued by the Mining Departments of the British Colonies of Australasia. I am largely indebted for special communications received from metallurgists in charge of prominent companies and important works, and from the officers of the government mining departments of the Australian Colonies. It has seemed advisable with some "improvements," and generally with the patents, to simply place them on record without any special comment.

Respectfully yours,

A. SCHEIDEL, Ph.D., E.M.

San Francisco, October 1, 1894.

PREFACE TO THIRD EDITION.

Hon. J. J. Crawford, State Mineralogist, San Francisco, Calif.:

Dear Sir: Following your instructions of March 3d, I herewith submit my notes referring to the progress the cyanide process and its practical application have made since the first edition of this monograph.

It has been my wish and intention to rewrite my paper and incorporate in it my experiences, and those of others, acquired during the last eighteen months; but I must regretfully abstain from this, in consequence of the limited time at my disposal.

The cyanide process, as described in my paper, has not undergone any revolutionary changes since I first reported upon it, and the chemical and mechanical methods remain, on the whole, unchanged. I have of late inspected a number of extensive and successful plants in different parts of the world, and I found the methods in use still the orthodox ones. The process is still chiefly applied to tailings, but its application to the direct treatment of ores is gradually extending.

The range of the process has not been considerably extended, although numerous attempts have been made for that purpose through the agency of added chemicals and by the chemical action exercised by the electric current.

Among the proposals for the improvement of the cyanide process may be mentioned the recommendation of the use of haloid compounds of cyanogen, particularly of bromide of cyanogen in presence of a cyanide solution, for the purpose of more quickly and more completely dissolving the gold. The practical efficiency of this process and its commercial value have yet to be demonstrated by exact experiments on many kinds of material, under varying conditions, and on a technical scale.

The application of oxidizing chemicals in connection with the cyanide solution for the purpose of dissolving gold more rapidly and more economically has not had that success which laboratory experiments with clean cyanide solutions acting on gold foil or leaf gold have led to anticipate; many experiments have proved that laboratory tests of that kind, although of considerable scientific value and interest, are no practical guidance for the treatment of ores and tailings.

Oxidizing action of chemical compounds in presence of cyanide solution can become a commercial success for gold extraction from ores only if thereby the consumption of cyanide is reduced, the percentage of extraction increased, and the time of operation diminished to such an extent as to compensate for the increased outlay on chemicals, or if any one of these conditions is achieved to such a degree as to offer commercial benefits beyond those produced by the application of cyanide alone and above the additional cost of the oxidizing agent.

More light is still wanting in reference to many secondary chemical reactions occurring in the technical application of the process and kindred matters, although valuable work has been done, chiefly referring to the action of oxygen on cyanide solutions,—the rate of solution of gold in cyanide solutions,—the selective action of weak cyanide solutions,—the action of carbonic acid on cyanide solutions,—the question is potassic zinc cyanide a solvent for gold,—the action of alkaline sulphides in cyanide solutions,—the detection of small quantities of potassium cyanate in potassium cyanide,—and the estimation of cyanogen in impure solutions. Much, however, remains to be done in many directions.

The technical methods of operation are generally still those of old; the appliances for percolation and agitation remain, on the whole, the same.

The difficulties in treating wet-crushed ores direct by cyanide have not been completely overcome, although great ingenuity and much mechanical skill have been brought to bear on the question. The object in view is to separate the slimes from the coarser particles to fit these for more uniform, more rapid, and more successful percolations. Two methods are being chiefly employed for the purpose of eliminating the slimes which are unavoidably present in wet-crushing: that of "direct filling" and that of "intermediate filling." That of "direct filling" consists in the hydraulic separation of the slimes from the coarse tailings by means of "Spitzlutten" and the direct delivery of the sand into the percolation-vats. By this method a second handling of the tailings is avoided and oxidation of pyritic matter is almost prevented. The other method in use—that of the "intermediate filling"—consists in the discharge of the tailings pulp, after coming from the stamp mill, by means of a rotating distributor made of pipes, attached to a stationary iron column, into an intermediate vat filled with water. The fine slime particles remain in suspension when the pulp flows into the vat, they overflow with the water, and are conveyed by means of launders into pits, where they finally settle. Wet-crushed ores can be handled by either method, like tailings.

Cyanide treatment of the slimes themselves is still one of the problems of the process, no cheap means having as yet been devised for their economical working. The question is of great commercial importance, in consequence of the large percentage of slimes produced in the wet battery process and the large amount of gold remaining unrecovered by their not being turned to account. The only process so far giving satisfactory results is that of treating the slimes by agitation and separating the gold solution and the washings by decantation.

The chief method for the precipitation of the gold from the cyanide solution is still that of zinc in form of threads or shavings, although electrical methods are now worked on a considerable scale.

Neither the cost of treatment nor the percentage of extraction has varied to any extent as compared with the figures given in the first edition. Slight improvements in the latter and slight reductions in the former have been made in individual cases.

The validity of the principal cyanide patent has been questioned in different countries, and judicial decisions have been given in reference; the matter, being still *sub judice* in some countries, is beyond the province of this paper.

The manufacture of cyanide of potassium has been further improved upon. The high-grade white material is more and more taking the place of the crude chemical which used to be the only reagent obtainable in large quantities in the market. The price of cyanide has not been reduced. A new source of cyanogen will be utilized in the gases of coke furnaces.

The literature on the process has been enriched by some valuable additions; several treatises on the application of the process in South Africa will be found of practical value. I refer to these for further special information on the question of the treatment of Rand ores and on all other matters in connection therewith, such as direct cyanide treatment without concentration and that of previous concentration and subsequent treatment of the concentrates and the tailings by themselves.

The number of works using the cyanide process for ore-treatment has considerably increased. The Rand gold-field is still easily leading in the quantity of gold produced by cyanide; the process has there become a factor of political economy of the first rank.

New Zealand was among the first countries to adopt the process, and a large number of cyanide plants are in successful operation in that colony. The introduction of the process into those districts where the fineness of the gold prevented the successful extraction of high percentages by other means has been of immense commercial importance, and an industry somewhat languishing before its introduction has reached a high state of development.

In Australia the process has not yet found that extensive application to be expected on a continent where the variety of ores and the quantity of valuable tailings are considerable; an increase in the use of the process is, however, noticeable. Several small plants have been recently erected on the newly discovered gold-fields of Western Australia for the purpose of treating dry-crushed ores direct. None of these have, as yet, realized expectations; the cause may be found partly in the adverse local conditions of scarcity and bad quality of water, and partly in the design of the plants, which did not sufficiently allow for these conditions and for the peculiarities of the ores.

The application of the process in the United States has been extended, but is still out of proportion to its possibilities.

A further development of the process and the solving of many of the problems which are still confronting the metallurgist will surely follow with the increasing number of extracting works.

The cyanide process has now taken the first position among hydro-metallurgical methods for gold extraction.

Respectfully yours,

A. SCHEIDEL, Ph.D., E.M.

Coolgardie, W. A., December 15, 1896.

THE CYANIDE PROCESS

ITS

PRACTICAL APPLICATION AND ECONOMICAL RESULTS.

By A. Scheidel, Ph.D., E.M.

The "cyanide process" for extracting gold and silver from ores is based on the fact that a diluted solution of potassium cyanide dissolves these metals, forming, respectively, auro-potassic cyanide and argento-potassic cyanide, from many ores, without dissolving to any material extent the other components thereof. The process consists of treating suitable ores, when finely divided, with a weak solution of potassium cyanide, either by allowing the solution to percolate through the ore or by agitating a mixture of the ore and solution. This part of the operation being completed, the solution is separated from the solid material and the gold and silver are precipitated in metallic form. This process for the extraction of gold and silver is comparatively old in its principle, but modern in its technical application. During the last four years it has been introduced into almost every gold field, and upwards of $14,000,000 in gold and silver have been recovered by the process, which demonstrates beyond doubt that it is one of the most important additions to the wet methods of gold and silver metallurgy. The aim of this paper is to present the history of the process and to describe the ores for which it is adapted, together with their preparation and manipulation during treatment. The economical features of the cyanide process are also dwelt on at some length. The text is illustrated by plans and diagrams.

The State Mining Bureau of California was among the first in the United States to investigate the merits of the cyanide process, as set forth in a paper by Dr. W. D. Johnston, in the Xth Report of the Bureau. The process has since found extensive application, and other valuable and interesting papers have been published, but an exact account of the methods employed in all parts of the world is still wanting. This writing is undertaken at the request of Hon. J. J. Crawford, State Mineralogist of California. The facts herein recorded are obtained from the practical experience of the writer in New Zealand and the United States, and of others who have been very successful in the application of the cyanide process. I take great pleasure in expressing my sincere appreciation of and gratitude for the assistance my contributors have extended. I desire especially to acknowledge my obligation to Mr. John S. MacArthur, of Glasgow, Scotland, and to Mr. J. M. Buckland, the general manager of the African Gold Recovery Company, Lim., in Johannesburg, South African Republic. In order to allow a comparatively full description of methods and appliances, in the following pages, theoretical matter is limited to the main chemical reactions incidental to the process, and an explanation of some of the difficulties most frequently met. To facilitate the consultation of the paper, I prefix the following synopsis:

I. History of process: Solubility of gold and silver in cyanide as known to Hagen, Bagration, Elsner, Faraday; its technical application by Wright and Elkington; its metallurgical application by Rae, Simpson, Endlich and Mühlenberger, Louis Janin, Jr., Dixon, MacArthur and Forrest, Molloy, A. Janin and Merrill, W. D. Johnston.
II. Scope of process.
III. Chemistry of process.
IV. Demonstration of the process. Methods of operation:

    A. The agitation process.
    B. The percolation process.
        (a) Percolation of ores.
        (b) Percolation of tailings.
        (c) Percolation of concentrates.
    C. Cyanide and cyanide solutions.
    D. Treatment of the gold solutions. Recovery of the gold and silver.
        (a) Precipitation by zinc.
        (b) The Molloy process.
        (c) The Siemens and Halske process.
        (d) The Pielsticker process.
        (e) The Moldenhauer process.
        (f) The Johnston process.

V. Percentage of extraction.
VI. Working costs of process.
VII. Cost of cyanide plants.
VIII. Machinery and appliances.
IX. Laboratory work.
X. Danger in working the process.
XI. Exemplification of the process. The process in various countries:

    A. Africa.
    B. Australasia.
        (a) New Zealand.
        (b) Tasmania.
        (c) Western Australia.
        (d) South Australia.
        (e) Queensland.
        (f) New South Wales.
        (g) Victoria.
    C. United States of America.
        (a) Utah.
        (b) Montana.
        (c) Colorado.
        (d) Nevada.
        (e) Arizona.
        (f) New Mexico.
        (g) South Dakota.
        (h) California.
    D. Mexico, Colombia, Straits Settlements, Russia, Borneo.

XII. Summary and conclusions.
XIII. Patents:

    Julio H. Rae. Improved mode of treating auriferous and argentiferous ores. U. S. patent 61,866, dated February 5, 1867.
    Thomas C. Clark. Extracting precious metals from ores. U. S. patent 229,586, dated July 6, 1880.
    Hiram W. Faucett. Process of treating ore. U. S. patent 236,424, dated January 11, 1881.
    John F. Sanders. Composition for dissolving the coating of gold in ore. U. S. patent 244,080, dated July 12, 1881.
    Jerome W. Simpson. Process of extracting gold, silver, and copper from their ores. U. S. patent 323,222, dated July 28, 1885.
    John Stewart MacArthur, Robert Wardrop Forrest, M.D., and William Forrest, M.B. Improvements in obtaining gold and silver from ores and other compounds. English patent 14,174; 1887.

XIII. Patents (*continued*):

John Stewart MacArthur, Robert Wardrop Forrest, and William Forrest. Process of obtaining gold and silver from ores. U. S. patent 403,202, dated May 14, 1889.

John Stewart MacArthur. Metallurgical filter. U. S. patent 418,138, dated December 24, 1889.

John Stewart MacArthur, Robert Wardrop Forrest, and William Forrest. Process of separating gold and silver from ores. U. S. patent 418,137, dated December 24, 1889.

Edward D. Kendall. Composition of matter for the extraction of gold and silver from ores. U. S. patent, dated September 13, 1892.

Bernard Charles Molloy. Improvements in precipitating and collecting metals from solutions containing them. English patent 3,024; 1892.

John Cunninghame Montgomerie. Improvements in the extraction of gold and silver from ores or compounds containing the same, and in apparatus applicable for use in the treatment of such materials by means of solvents. English patent 12,641; 1892.

John Stewart MacArthur and Charles James Ellis. Improvements in extracting gold and silver from ores and the like. New Zealand patent-specification, June 29, 1893.

Carl Moldenhauer. Improvements in recovering gold and other precious metals from their ores. New Zealand patent-specification, August 31, 1893.

Carl Pielsticker. Improvements in the extraction of gold and silver from ores. New Zealand patent-specification, December 14, 1893.

Alexis Janin and Charles W. Merrill. Process of leaching ores with solutions of alkaline cyanides. U. S. patent 515,148, dated February 20, 1894.

William David Johnston. Method of abstracting gold and silver from their solutions in potassium cyanides. U. S. patent 522,260, dated July 3, 1894.

XIV. List of plans, diagrams, and tables:

Details of the false bottoms of the percolation vats. (W. R. Feldtmann.)
Plant to treat a minimum of 2,000 tons per month. (MacArthur.)
Table giving sizes and material of percolation vats. (A. Scheidel.)
Zinc box. (MacArthur.)
Zinc filter. (A. Scheidel.)
Porcelain filter. (A. B. Paul.)
Table giving extraction results on various ores. (A. Scheidel.)
Discharging tailings-vats at the Langlaagte Estate Company's plant. (W. R. Feldtmann.)
Square filter vats at the works of the Crown Company, with doors for the discharging trucks. (Irvine.)
Variation No. 1 in designs of cyanide plants. (W. R. Feldtmann.)
Variation No. 2 in designs of cyanide plants. (W. R. Feldtmann.)
Variation No. 3 in designs of cyanide plants. (W. R. Feldtmann.)
Side discharge at percolation vats. (W. R. Feldtmann.)
Bottom discharge at percolation vats. (Chas. Butters.)
Bottom discharge at percolation vats. (W. E. Irvine.)
The cyanide works of the Robinson Company. (Chas. Butters.)
The cyanide plant of the Crown Company. (John MacConnell.)
The Sylvia Company's Cyanide Works, Tararu, Thames, New Zealand. (A. Scheidel.)
Melting room for cyanide bullion in the Sylvia Company's Works, Tararu, New Zealand. (A. Scheidel.)
The tailings cyanide works at Waihi. (A. James.)
The concentrating and cyanide extraction works of the Sylvia Gold and Silver Mining Company. (A. Scheidel.)
The cyanide plant of the Sylvia Company (plan and longitudinal section). (A. Scheidel.)
Table showing the concentrating and cyanide process in the works of the Sylvia Company. (A. Scheidel.)
Bullion furnace. (A. Scheidel.)
Wet-mill cyanide plant, Revenue. (F. B. & R. B. Turner.)
Dry-mill cyanide plant, Revenue. (F. B. & R. B. Turner.)
Utica Company cyanide plant (plan and longitudinal section). (A. Scheidel.)
The vacuum filter, Utica cyanide plant. (A. Scheidel.)
The bullion filter, Utica cyanide plant. (A. Scheidel.)
Agitator. (J. H. Rao.)
Metallurgical filter. (J. S. MacArthur.)
Apparatus for treatment of ores, etc., by means of solvents. (J. C. Montgomerie.)
Improved apparatus for the extraction of gold and silver from ores. (C. M. Pielsticker.)
Table giving analysis of gold production in the Witwatersrand District for April, 1894. (Witwatersrand Chamber of Mines.)

XV.   List of abbreviations of literature:

E. & M. J.—Engineering and Mining Journal, New York.
M. I.—Mineral Industry.
M. S. P.—Mining and Scientific Press.
Tr. A. I. M. E.—Transactions of the American Institute of Mining Engineers.
J. S. Chem. I.—Journal of Society of Chemical Industry, England.
J. fr. Chem.—Journal für practische Chemie.
J. Ch. S.—Journal Chemical Society.
Tr. Phil. Soc.—Transactions of the Philosophical Society.
M. Sc.—Moniteur Scientifique.
A. Ch. Ph.—Annales de Chimie et de Physique.
Ch. N.—Chemical News.
B. A. I. Sc.—Bulletin de l'Académie Imperiale des Sciences de St. Petersbourg.
B. S. Ch.—Bulletin de la Société Chimique de Paris.

I. HISTORY.

The fact of gold being soluble in cyanide of potassium solution has been known for a considerable time. Hagen is reported to have mentioned it in 1806. Dr. Wright, of Birmingham, England, used gold-cyanide solution for electroplating in 1840; he made this application in consequence of his studies of Scheele's report on the solubility of gold-cyanide in a cyanide of potassium solution. J. R. & H. Elkington patented Wright's invention; they speak in their patent-specification of a boiling solution of gold or cyanide of gold in prussiate of potash. The first record in scientific literature of experiments in which metallic gold was dissolved in a cyanide of potassium solution consists in Prince Pierre Bagration's paper in the Bulletin de l'Académie Imperiale des Sciences de St. Petersbourg, 1843, t. 11, p. 136. Bagration, who alludes to Elkington's process, preserved cyanide of potassium solution in a dish, gilded on the inside. He noticed that after eight days the whole gold surface had been attacked. He experimented then with finely divided gold under the influence of the galvanic current; the latter he soon recognized as not of any benefit in the dissolving process. He precipitated the gold out of the cyanide solution by means of the electric current on a cathode of copper. Continued experiments proved the advantage of higher temperature during the dissolving process, and taught the precipitation of gold from its still warm solution by means of silver or copper plates, without the electric current. The higher temperature had, however, the disadvantage of the silver and copper being strongly attacked by the cyanide solution during the precipitation process. Bagration extended his experiments to solutions of ferrocyanide, which he found to act like cyanide, but in a much less degree. He further studied the solubility of gold in the form of plates, in cyanide, and found it to be dissolved in such form at a considerable rate at a temperature of 30° to 40° C. He noticed the influence of the air on the reaction. Bagration believes that hydrocyanic acid in a state of generation is a gold solvent, and he concludes his paper with the remark that in the future, cyanide of potassium must be enumerated among the solvents of gold. L. Elsner published in J. fr. Chem., 1844, p. 441, his observations on the reactions of "reguline metals" in an aqueous solution of cyanide. He found that gold and silver were dissolved in potassium cyanide without decomposition of water. "The dissolution of the metals is, however, the consequence of the action of oxygen, which, absorbed from the air, decomposes part of the cyanide." His reaction has been expressed by others in the following equation:

$$2\,\mathrm{Au} + 4\,\mathrm{KCy} + \mathrm{O} + \mathrm{H_2O} = 2\,\mathrm{AuKCy_2} + 2\,\mathrm{KOH}$$

(Gold.)   (Cyanide   (Oxygen.) (Water.)   (Auro-potassic   (Potassic
       of potassium.)   cyanide.)   hydrate.)

It is generally called Elsner's equation. Some years after, Faraday made use of the solubility of gold in cyanide solution for reducing the thickness of gold films (Exp. relations of gold and other metals to light, Tr. Phil. Soc., 1857, p. 147). The basis of the most modern process for the extraction of gold was thus provided. It took many years, however, before the enumerated facts were made use of for the extraction of gold from ores. In 1867, Julio H. Rae took out United States patent No. 61,866, dated February 5th, for an "improved method of treating auriferous and argentiferous ores" with a current of electricity in connection with suitable liquids—such, for instance, as cyanide of potassium. Rae's process is an agitation process; he proposed to "expose the auriferous or argentiferous rock to the combined action of a current of electricity and of suitable solvents, and to separate the gold or silver from the rocks containing the same by the action or aid of electricity." The principle of Rae's process, as stated by him, distinguishes his method from the modern cyanide process. His method does not appear to have advanced beyond the laboratory stage or to have found extensive and successful practical application, and it sank into oblivion. Since then, cyanide of potassium in connection with gold and silver metallurgy has repeatedly been made a patent claim; in many cases, however, the application recommended is in its principle different from the application which characterizes the modern cyanide process. Thomas C. Clark, of Oakland, Cal. (United States patent No. 229,586, July 6, 1880), roasted his ore to a red heat, and placed it in that condition in a cold bath composed of a solution of salt, prussiate of potash, and caustic soda. H. W. Faucett, of St. Louis, Mo. (United States patent No. 236,424, January 11, 1881), subjects hot crushed ores to the action of disintegrating chemicals, cyanide of sodium among others, in solution under pressure, the pressure being effected by the steam generated by the contact of the hot ores with the chemical solution in a closed vessel. This treatment, like that proposed by Clark, was intended as preliminary to amalgamation. John F. Sanders, of Ogden, obtained United States patent No. 244,080, dated July 12, 1881, for "composition for dissolving the coating of gold in ore." This composition is made of cyanide of potassium and glacial phosphoric acid. He stated that by using this mixture he could dissolve "the impure coatings of gold, leaving the gold free and exposed, and permitting it to be amalgamated." It is evident, therefore, that these processes bear no similarity or relation to the modern cyanide process. For a considerable time, cyanide of potassium has been used in the gold fields of California and Australasia for removing film-coating from gold in ores; its application in the pan-amalgamation process may have been a source of loss of gold.

The application of a cyanide of potassium solution for the extraction of gold and silver direct from their ores, which application had been neglected since Rae, was taken up again by Jerome W. Simpson, of Newark, N. J., who obtained United States patent No. 323,222, dated July 28, 1885, for a process of extracting gold, silver, and copper from their ores. Simpson reduced his ore to a powder and agitated it with a solution of certain salts, which combine chemically with the metal in the ore and form therewith a soluble salt. The salt solution was composed of one pound of cyanide of potassium, one ounce of carbonate of ammonia, one half ounce of chloride of sodium, and sixteen quarts of water. This solution is described as particularly adapted to ores containing gold, silver, and copper in the form of sulphurets. In the absence of silver no chloride of sodium is used; for ores rich in silver a proportionately larger quantity of chloride of sodium is employed. The metals dissolved in the salt solution were precipitated by means of zinc, suspended therein in form of pieces or plates. Simpson was aware that cyanide of potassium, in connection with an electric current, had been used for dissolving metal, and also that zinc had been employed as a precipitant. What he claims as new is: (1) The process of separating gold and silver from their ores, which consists in subjecting the ore to the action of the solution of cyanide of potassium and carbonate of ammonia, and subsequently precipitating the dissolved metal by means of zinc. (2) The process of separating metals from their ores, to wit, "subjecting the ore to the action of a solution of cyanide of potassium, carbonate of ammonia, and chloride of sodium, and subsequently precipitating the dissolved metals."

My own experiments have proved that the addition of sodium chloride is of no benefit for the extraction of silver. The addition of ammonium carbonate is not beneficial to the extraction of either gold or silver, except under certain conditions, when it may be substituted advantageously by an alkali or an alkaline earth; in the presence of base metals it is of disadvantage. Simpson's patent description appears to indicate that he had not discovered the most important property of dilute cyanide solutions, namely, that of dissolving, without the addition of other chemicals, the noble (in preference to the base) metals. His patent-claim consists eminently in adding to the cyanide solution the chemicals mentioned above. His process, like that of Rae, is an agitation process. The zinc for precipitating the bullion he used in the form of plates or pieces. There is no record in the technical literature in reference to the application of Simpson's process before the issue of the MacArthur and Forrest patents in 1889. After Rae and Simpson, others have made experiments with cyanide solutions for the purpose of gold and silver extraction from ores. F. M. Endlich and N. H. Mühlenberger are reported to have filed a caveat in 1885, without, however, securing a patent, the former having apparently become doubtful as to the applicability of cyanide as an economical process (E. & M. J., 1891, p. 86). Louis Janin, Jr., made interesting experiments in the same direction in Park City, Utah, of which he published the results in 1888 (E. & M. J., 1888, p. 548). These experiments refer chiefly to silver extraction, but he mentions as well the results on gold ores; his results appear to have been encouraging, and led to his filing a caveat on May 1, 1886, but this was not pushed on to the taking out of a patent.

In the southern hemisphere, W. A. Dixon has made experiments with cyanide on Australian ores as early as 1887. He recorded his results, which are at least of historical interest, in a paper read before the Royal Society of New South Wales. Dixon describes therein the experiments made by him at the instigation of the Government Inspector of Mines, who suggested that the extraction of gold from complex minerals was a subject worthy of investigation. Dixon tried on such ores amalgamation and a number of solvents. He found "the aurocyanides of the alkaline metals of sufficient stability to render their use possible for the extraction of the gold." He mentions Bagration's and Elsner's publications and alludes to Rae's patent, of which, however, he possessed no particulars. Dixon feared that "the high price of cyanide, its instability when exposed to the air, and its extremely poisonous qualities," would prove such obstacles as to preclude its use for metallurgical purposes. He found the reaction between gold and cyanide slow if "the gold was at all dense"; in presence of alkaline oxidizing agents, however, he found the dissolving process sufficiently rapid. Dixon experimented also with the ferrocyanide of potassium. His results generally, did not, as far as known, lead to the metallurgical application of cyanide as a gold and silver solvent.

I have thus described the history of cyanide of potassium as a gold dissolving agent from the early laboratory experiments up to its metallurgical application for ore extraction; this latter, however, did not gain any practical importance until John S. MacArthur and W. Forrest, of Glasgow, Scotland, took out their patents for the use of cyanide as a gold and silver solvent from ores, and gave thereby the cyanide process a start all over the world. Their patents mark an epoch in gold metallurgy. The results of the application of cyanide, as suggested by them, have been very satisfactory; the $14,000,000 of bullion produced by it during the five short years of its working represents what, by the ordinary methods, would have been irrecoverably lost; hence its value and importance from the standpoint of metallurgy and political economy. The experiments of MacArthur and Forrest with gold dissolving reagents occupied some years before their English cyanide patents were applied for (J. S. MacArthur, J. S. Chem. I., March 31, 1890, No. 3, vol. 9). They drew out a list of possible solvents having a stronger affinity for gold than for sulphides, which included the cyanides, and which they found to solve the problem. Their experiments, conducted first on a small scale and with ores of many kinds and of different sources, were so satisfactory that they gradually worked on a larger scale, and their results formed the basis for the introduction of the cyanide process into most gold-producing fields. Their English patent was applied for October 19, 1887. Since then they applied for and obtained patents in many gold-producing countries. Their United States patents are dated as follows: 403,202, May 14, 1889; 418,137, December 24, 1889; 418,138, December 24, 1889. Their invention is described " as having principally for its object the obtaining of gold from ores, but it is also applicable for obtaining silver from ores containing it whether with or without gold, and it comprises an improved process, which, while applicable to auriferous and argentiferous ores generally, is advantageously and economically effective with refractory ores, or ores from which gold and silver have not been satisfactorily or profitably obtainable by the amalgamating or other processes hitherto employed; such as ores containing sulphides, arsenides, tellurides, and compounds of base metals generally, and ores from which the gold has not been easily or completely separable on account of its existing in the ores in a state of extremely fine division." The patentees describe their invention (I am following United States patent 403,202) as consisting in subjecting the ores to the action of a solution containing a small quantity of cyanide, without any other chemically active agent. In dealing with ores containing per ton twenty ounces or less of gold or silver, or gold and silver, they find it most advantageous to use a quantity of cyanide, the cyanogen of which is equal in weight to from one to four parts for every thousand parts of the ore dissolved in a quantity of water of about half the weight of the ore; they generally use a solution containing two parts of cyanogen for every thousand parts of the ore. In the case of richer ores, while increasing the quantity of cyanide to suit the greater quantity of gold or silver, they also increase the quantity of water so as to keep the solution dilute; in other words, the cyanide solution should contain from two to eight parts, by weight, of cyanogen to one thousand parts of water, and the quantity of the solution used should be determined by the richness of the ore. The patentees state: " By treating the ores with the dilute and simple solution of a cyanide, the gold or silver is, or the gold and silver are, obtained in solution, while any base metals in the ores are left undissolved except to a practically inappreciable extent; whereas, when the cyanide is used in combination with an electric current, or in conjunction with another chemically active agent, such as carbonate of ammonia, or chloride of sodium, or phosphoric acid, or when the solution contains too much cyanide, not only is there a greater expenditure of chemicals in the first instance, but the base metals are dissolved to a large extent along with the gold or silver, and their subsequent separation involves extra expense, which is saved by their process." Later on MacArthur and Forrest obtained patents covering the use of zinc in a fine state of division for the purpose of precipitating gold and silver from cyanide, chloride, bromide, thiosulphide, sulphate, or other similar solutions; they further protected the use of an alkali or alkaline earth for neutralizing ores preparatory to subjecting the same to the action of cyanogen or of a cyanide. The MacArthur-Forrest patent-claims consist, therefore, in three points: (1) The application of diluted solutions of cyanide (not exceeding eight parts of cyanogen to one thousand parts of water); (2) the use of zinc in a fine state of division; (3) the preparatory treatment of the ore, which has become partially oxidized by exposure to the weather, with an alkali or alkaline earth, for the purpose of neutralizing the salts of iron or other objectionable ingredients formed by partial oxidation.

It is not the purpose of this paper, which is intended to describe the historical development of the cyanide process and its present forms of application, to enter into a judicial discussion of patent-claims and patent-rights. It is the duty of the historian to date the cyanide process as a commercial success from 1890, when it was introduced as "the MacArthur-Forrest process" on the Witwatersrand gold fields, in the South African Republic. Its success as a metallurgical experiment may be dated from the tests made on a large scale with ore from the New Zealand Crown Mine in June and July, 1888. The practicability of the cyanide process once established, others endeavored to introduce improvements in its application, which they protected by letters patent. A patent which once promised to become of practical importance is that of B. C. Molloy, of Johannesburg, whose "improvement" consists in the abolition of zinc as a precipitant of gold, and in the revivification of the cyanide of potassium in the solution. In this process the ore is treated with cyanide of potassium as usual; the resulting liquors are passed through a "patent Molloy separator," which consists of an amalgamator, the mercury of which is constantly being charged electrolytically with potassium. The potassium on coming into contact with the water of the solution decomposes it with the evolution of hydrogen and the formation of the oxide of the alkaline metal. The nascent hydrogen decomposes the solution of the cyanide of gold, and sets the gold free, which is precipitated upon and collected by the mercury; the metal of the alkaline oxide reacts upon the cyanogen compound, and so reproduces the cyanide of potassium. The original solution, thus regenerated, is then ready for use again. (In reference to further details see under precipitation of gold and silver, p. 38.)

Among other cyanide patent-specifications may be mentioned the following: John C. Montgomerie, of Scotland, obtained English patent No. 12,641, 1892, "for improvements in the extraction of gold and silver from ores and in apparatus applicable for use in the treatment of such by means of solvents." His process is the well-known agitation process of finely divided ore with cyanide solution and the addition of an alkaline oxide "for the purpose of economizing the solvent and expediting its action." The patent-specification does not contain any claims which might be termed either an invention or an improvement, either chemically or mechanically. One of the latest additions to the cyanide patent literature is United States patent 515,148, dated February 20, 1894, of Alexis Janin and Charles W. Merrill, for a process of leaching ores with solutions of alkaline cyanides. (For patent-specification, see Appendix.) They claim as new "the art of leaching ores with solution of alkaline cyanides, which consists in first leaching the ore with such solutions, then adding to the solution an agent which will precipitate the silver present as a sulphide, and then precipitating the gold in the solution with metallic zinc." The practical advantages of this complication will have to be proved.

A patent description of interest, although not a "cyanide process" strictly speaking, is that of E. D. Kendall, of Brooklyn, N. Y., dated September 13, 1892, who claims the use of potassium ferrocyanide combined with cyanide of potassium, for extracting gold and silver from ores, etc., as his invention. (For patent-specification, see Appendix.)

A further addition to the patent literature is the specification of MacArthur and Ellis, who propose to increase the efficiency and economy of the process in cases in which from the nature of the ores treated or other circumstances, soluble sulphides are formed, which retard and objectionally affect the action of the cyanide on the precious metals by adding to the ore or the cyanide solution suitable salts or compounds of metals which will form with the sulphur of the soluble sulphides an insoluble or inert sulphide. For this purpose preference is being given to the metallic salts or compounds in the following order: Salts or compounds of lead—such as plumbates, carbonates, acetate or sulphate of lead—sulphate or chloride of manganese, zincates, oxides, or chloride of mercury, ferric hydrate or oxide. The proportion to be used is easily to be ascertained by trials of a few samples in each case. (See patent-specification of John Stewart MacArthur and Charles T. Ellis, in Appendix.)

C. Moldenhauer proposes to render the cyanide process more expeditious and considerably cheaper by, firstly, adding to the cyanide solution an artificial oxidizing agent, by preference ferricyanide of potassium in alkaline solution, and, secondly, in precipitating the extracted precious metal out of its cyanide solution by means of aluminium, or alloys, or amalgam thereof. (See patent-specification in Appendix.)

C. M. Pielsticker reverts to the application of the electric current, in conjunction with the cyanide solution. He proposes to continuously circulate the solvent, to continuously precipitate the dissolved precious metals by electrolysis, and continuously regenerate thereby the reagent. (See patent specification in Appendix.)

The latest patent in connection with cyanide treatment of ores is that of Dr. W. D. Johnston, "for abstracting gold and silver from their cyanide solutions by means of pulverized carbon" (for further details, see page 40).

To make this report as complete as circumstances permit, I append the specifications of the patents which have been mentioned in the body of this paper, that the mining public may know the exact wording of descriptions and claims.

Such is the history of the cyanide process, rapidly sketched by tracing its development through the phases of its evolution and the intricacies of its patent literature. The modern cyanide process consists in the treatment of ores by means of dilute cyanide of potassium solutions, as a rule without the addition of other chemical substances, and in the subsequent precipitation of the gold and silver from the solution by means of zinc in form of shavings. It is commonly known as the Mac-Arthur-Forrest process. I now propose to enter into a description of the process itself. I embody in it the information given me by Mr. John S. MacArthur, of Glasgow, Scotland.

II. SCOPE OF PROCESS.

The process can be advantageously applied to many gold ores and many silver ores, and is often suitable for ores which are generally considered as rebellious or refractory. The word "ore" is here meant to include ores, tailings, concentrates, and all similar products from ore. The term "refractory" is used to signify any ore which cannot be satisfactorily amalgamated. The refractory character of such ores can be caused by the presence of base metals in combination with sulphur or arsenic, or otherwise by their physical structure, which prevents the gold from coming in contact with the mercury during the amalgamation process. To the latter class belong the ores in which the gold is "coated" with substances which prevent metallic contact ("rusty gold"). (An excellent instance of such coated ore is that found in the Mount Morgan Mine, in Queensland, where the finely divided gold is coated with a film of what has been termed hydrous peroxide of iron, which makes the gold absolutely refractory to amalgamation.) To the same class of refractory ores belong those in which the gold is so finely divided that the film of air surrounding the auriferous particles prevents amalgamation even under the most favorable conditions. The base metals which most frequently accompany refractory ores are iron, zinc, lead, copper, and antimony—usually as sulphides, sometimes as arsenides. When ores containing gold, silver, copper, zinc, iron, etc., are treated with solutions of cyanide of potassium, these metals are dissolved more or less, forming soluble cyanides. The solvent action on the base metals can be reduced to a minimum by reducing the strength of the solutions, the readily soluble gold and silver being easily dissolved out with only traces of copper, zinc, etc. The action of these weak cyanide solutions on iron, lead, arsenic, etc., is practically *nil*, and the solvent action on copper or zinc depends much upon the state of chemical combination in which they exist.

The cyanide process is adapted to treat most of such refractory ores as are described above. The principal exceptions are the ores which contain hydrated copper oxides and copper carbonates, and those which contain an appreciable quantity of antimony. "When copper compounds exist in a state physically hard, the cyanide solution does not readily act on them; but when the copper compounds are soft, porous, and spongy, the action of the cyanide is so decided as to interfere materially with its action on gold" (MacArthur). In reference to copper sulphide I found it no impediment to the process; carbonate of copper, however, was so readily attacked by cyanide that its presence proved absolutely prohibitive to the extraction of silver and interfered seriously with the extraction of gold. This most refractory ore, that I am speaking of, came from old workings in the Sylvia Mine, Tararu, New Zealand, where part of the ledge containing a large percentage of copper pyrites had been exposed for many years to the influence of moisture and the atmosphere; the resulting carbonate was hard, but notwithstanding this its reaction on cyanide solutions was very marked. One and a fourth ounce of such copper ore, finely divided and shaken for less than fifteen minutes with a 2.73 per cent cyanide of potassium solution, reduced the strength of the solution to 0.05 per cent of cyanide. The treatment of the ore in question proved that the affinity of cyanide to gold is at least equal to that of cyanide to copper, and very much greater than to silver, as, notwithstanding the rapid consumption of cyanide by the copper compound, upwards of 70 per cent of the gold assay-value was extracted by cyanide solution of the usual strength, whereas at the same time absolutely no silver had gone into solution. A preliminary treatment of such ore by sulphuric acid had a beneficial effect on the consumption of cyanide and thereby on the extraction of silver.

"In the case of antimonial ores, there is little or no interaction between the antimony and the cyanide, consequently the latter is not taken up; but as gold seems to be very firmly held by antimony, and as the compound is very impervious, the cyanide is unable to penetrate the mass, and to dissolve and separate the precious from the base metals. In the case of both copper and antimony the cyanide solution will act, but in the case of copper, if there is much present and acted upon, the consumption of cyanide is so great that the operation is not profitable, and in the case of the antimonial ores, though the cyanide will act with fine grinding and long contact, the expense involved often overbalances the value of the gold contents" (MacArthur). The physical state in which obnoxious compounds are found, is of the greatest importance. Hard-surfaced crystals are, even if finely divided, naturally less acted upon by cyanide than soft, spongy masses of the same size. For technical purposes, cyanide treatment of any ore will be called unsuccessful if the large consumption of cyanide precludes a commercial success, although finally a satisfactory extraction in percentage may be achieved.

III. THE CHEMISTRY OF THE PROCESS.

The chemical reaction on which the cyanide process of gold extraction rests is that of the formation of the double cyanide of gold and potassium:

$$2\,Au + 4\,KCy + O + H_2O = 2\,AuKCy_2 + 2\,KOH$$

(Gold.)   (Cyanide of potassium.)   (Oxygen.)   (Water.)   (Auro-potassic cyanide.)   (Potassic hydrate.)

That of silver extraction produces the double cyanide of silver and potassium:

$$2\,Ag + 4\,KCy + O + H_2O = 2\,AgKCy_2 + 2\,KOH$$

(Silver.)   (Cyanide of potassium.)   (Oxygen.)   (Water.)   (Argento-potassic cyanide)   (Potassic hydrate.)

Silver in the metallic state is, however, rarely met with in ores which are subjected to cyanide treatment. The part taken by oxygen in these reactions, first noticed by Prince Bagration and later confirmed by Elsner, has of late been disputed, but again confirmed by McLaurin, who published his experiments (J. Ch. S., 1893, May, p. 724) in reference to the question, and came to the following conclusions: (1) That oxygen is necessary for the solution of gold in cyanide of potassium, and that it combines with the potassium of the potassium cyanide in the proportions required by Elsner's equation; (2) That the rate of solution of gold in a solution of potassium cyanide passes through a maximum in passing from dilute to concentrated solution, and this remarkable variation is capable of explanation by the fact that the solubility of oxygen in a cyanide solution decreases with the concentration. The double compounds of cyanide of potassium and gold and silver, respectively, have been described in the Annales de Chimie et de Physique, 53, p. 462, 1858, and in Bull. de la Société Chimique de Paris, 29, 1878, p. 460. Both compounds are easily soluble in water.

The cyanide process, as illustrated by the before-mentioned equations, appears very simple indeed. Its adoption in many places has been very rapid, and its success, particularly on the tailings of the Johannesburg mills, has been great. The practical working and technically successful carrying out of cyanide treatment of any ore, even under the most favorable circumstances, is beset with complications, which require a careful study of all the circumstances connected with the case. All operations offer occasions for loss and opportunities for improvement. The reaction between cyanide and the metals, so simple in theory, is in practice more or less complicated by the reaction of other ore compounds on the cyanide and by other causes which it will be useful to investigate. That such reactions take place is put in strong evidence by the amount of cyanide consumed in treating ores, which is always considerably larger than the quantity theoretically necessary to dissolve the gold. In accordance with Elsner's equation, 10 parts of cyanide should dissolve 15.12 parts of gold. In the works at Johannesburg, however, in treating free-milling ore, 40 parts of cyanide are required to dissolve one part of gold; that is to say, 40 parts of cyanide are consumed for each part of gold obtained. The main reason for this fact must be looked for in secondary reactions, which as yet have only been partly studied. The great loss of cyanide takes place during the extraction process, and particularly during the first part of it, as proved by the rapid diminution in the strength of the solution. The loss of cyanide in the zinc boxes has often been exaggerated (see page 34). A loss of cyanide occurs by absorption in vats and tanks, which is given as high as one pound per ton of ore in Johannesburg (Butters and Clennell). Some loss will always result from the action of carbonic acid gas, which is always present in the atmosphere, and displaces cyanogen from the alkali, setting prussic acid free, which escapes into the air; if caustic alkali is present the freed prussic acid will be neutralized. The extent of loss by hydrolysis requires further investigation. The presence of free sulphuric acid or other products of more or less advanced decomposition of pyritic matter will naturally considerably interfere with the simple reaction by increasing the consumption of cyanide, and may, under the most unfavorable circumstances, completely prevent successful treatment. " In many cases tailings which

2cp

have been exposed to the weather contain oxidized compounds, such as sulphate of iron, and similar sulphates of alumina and magnesia, formed by the action of the metallic sulphates on the earthy constituents of the ore." When this is the case it is advisable to give such tailings one or more preliminary water-washings, because the cyanide is partly absorbed and partly decomposed by these substances, as seen in the following equations:

$$FeSO_4 + 2\,KCy = FeCy_2 + K_2SO_4$$
$$Fe_23(SO_4) + 6\,KCy + 3\,H_2O = Fe_2O_3 + 3(K_2SO_4) + 6\,HCy$$

From these equations it will be seen that the ferrous oxide combines with cyanogen, and that the sulphuric acid, forming the second constituent of the ferric salt, liberates hydrocyanic acid, which being volatile is not available, and moreover constitutes a loss and a danger. The action of the sulphates of alumina and magnesia has not been generally and sufficiently recognized. These salts act practically as if they were sulphuric acid: hydrocyanic acid is liberated and alumina or magnesia, as the case may be, precipitated, as shown in the following equations:

$$Al_23(SO_4) + 6\,KCy + 3\,H_2O = Al_2O_3 + 3\,(K_2SO_4) + 6\,HCy$$
$$MgSO_4 + 2\,KCy + H_2O = MgO + K_2SO_4 + 2\,HCy$$

The remedy for these troubles is, as before stated, water-washing, in some cases followed by a lime or soda treatment. Reference only has been made to ferrous sulphate as a soluble salt, but it has been found that the basic ferrous salts, which exist to a greater or less extent in "weathered" tailings, are insoluble in water, and yet act detrimentally on cyanide. In any case it is difficult to wash out the last traces of any soluble substance, and it is wise to economize cyanide by an alkaline treatment. "While ferrous salts, soluble or insoluble, exist in the tailings, the lime or soda combines with the acid and deposits the ferrous oxide or hydrate in the tailings. The ferrous oxide would still absorb cyanogen if a cyanide solution were present, but if the air has free access before a cyanide solution is applied, the ferrous oxide is oxidized to ferric oxide, which does not combine directly with cyanogen. It will thus be seen that where salts of iron have to be dealt with it is advisable to make the alkaline treatment preliminary to permit of the necessary oxidation; but where sulphates of alkaline earths only are in question, the requisite lime or alkali may be added along with the cyanide solution. Where soluble iron salts are present to any extent, the washing should be very thorough, and the solution should be run off from the vat through a separate pipe which has no connection with any of the cyanide pipes" (or better, the washing should take place in a special vat; see page 49). "This matter of salts formed by oxidation arises chiefly in the case of tailings, but it may also happen with concentrates and ores, in which case they are treated as tailings" (MacArthur).

Butters and Clennell advance the following equations of possible reactions accompanying the action of cyanide on pyrites. They illustrate first the influence of oxygen on pyrites:

$$FeS_2 + H_2O + 7O = FeSO_4 + H_2SO_4$$

$$2 FeSO_4 + O = Fe_2O_3, 2 SO_3 \text{ (Wittstein)}$$

$$10 FeSO_4 + 5 O = \underset{\substack{\text{(Basic sulphate}\\\text{insoluble.)}}}{2 Fe_2O_3, 2 SO_3} + \underset{\substack{\text{(Ferric sulphate}\\\text{soluble.)}}}{3 Fe_2(SO_4)_3} \text{ (Berzelius)}$$

They describe, then, the reaction of cyanide on such products:

$$FeSO_4 + 2 KCy = FeCy_2 + K_2SO_4$$
$$FeCy_2 + 4 KCy = K_4FeCy_6$$

Ultimately giving rise to

$$3 K_4FeCy_6 + 6 FeSO_4 + 3 O = Fe_2O_3 + 6 K_2SO_4 + Fe_7Cy_{18}$$

Ferric salts and cyanide give:

$$Fe_2(SO_4)_3 + 6 KCy = Fe_2Cy_6 + 3 K_2SO_4$$

and,

$$Fe_2Cy_6 + 6 H_2O = Fe_2(OH)_6 + 6 HCy$$
$$Fe_2(SO_4)_3 + 6 KCy + 6 H_2O = Fe_2(OH)_6 + 6 HCy + 3 K_2SO_4$$

A mixture of ferrous and ferric sulphates on addition of cyanide will form Prussian blue when the ferric salt is in excess:

$$18 KCy + 3 FeSO_4 + 2 Fe_2(SO_4)_3 = 9 K_2SO_4 + Fe_4(FeCy_6)_3$$

and Turnbull's blue when ferrous salt is in excess:

$$12 KCy + 3 FeSO_4 + Fe_2(SO_4)_3 = 6 K_2SO_4 + Fe_3(FeCy_6)_2$$

The reactions between the various iron and cyanogen compounds are very complicated, and a number of possible reactions have been illustrated by equations by various writers, the discussion of which here would take up too much space.

In cases where such conditions exist, a preliminary washing with water alone, or with solutions of carbonates or hydroxide of sodium or lime, as described, may be not only useful but imperative. A great surplus of alkali should be avoided, on account of its action on the zinc in the precipitation boxes. The loss of zinc will be larger the greater the alkalinity of the solution; besides this, it is apt to form a sulphide of sodium or potassium with the sulphur of ores, which interferes with the extraction of the silver. Further careful scientific researches in reference to secondary reactions, which accompany the cyanide process, will probably lead to technically important results. The chemistry of precipitating the metals from cyanide solution will be discussed in connection with the description of the various methods employed for that purpose.

IV. DEMONSTRATION OF THE PROCESS — METHODS OF OPERATION.

The cyanide process is worked either by agitating the ore with the solution ("the agitation process"), or by allowing the solution to pass through the ore ("the percolation process").

A. The Agitation Process.

When cyanide treatment of ores was first attempted, it was done by agitating the material under treatment with cyanide solution; Rae's cyanide process of 1867 and Simpson's process of 1885 were agitation processes. Generally speaking, agitation, as compared with percolation, expedites and in instances increases extraction, but it requires motive power, which is a source of expense. Wherever large quantities of ore are being treated it has been abandoned in favor of the percolation process. It is useful, however, in many instances where the ores are hard and dense, and of a sufficient high value to pay for the necessary motive power and permit a convenient method of filtration; it is applied where the quantities are limited and is mostly used for treating concentrates, or such ores as make the treatment of limited parcels by themselves desirable. The importance of the cyanide agitation process has not been so fully recognized as, in some instances, it deserves. It is natural that if percolation gives as cheaply the same results it will be preferred, but sometimes the agitation system has the advantage of giving quicker, higher, and cheaper returns. Some ores, particularly ores containing tellurides and sulphide of silver, give better results by agitation than by percolation. The agitation process, in its present form, is not well adapted to handling very large quantities of ore without a considerable outlay of machinery. Technical improvements of the system, which in suitable cases may make the whole process almost a continuous one, may be expected. The chief appliances for the agitation process are the agitator and the filter. Although any vat fitted with revolving arms and barrels, similar to those employed in chlorination, may be used successfully for agitation, still an agitator which permits a charge and discharge quickly and safely, which has the least wear and tear, does absorb neither gold nor cyanide, and is cheap in its first cost, corresponds best with all requirements. I have been using wooden barrels, wooden vertical agitators, iron pans, and steel cylindrical agitators, and have found the latter construction best suited to the purpose and satisfying all the above conditions. (For description of such an agitator, see Utica cyanide plant, page 89.)

For the purpose of extraction, the ore and the cyanide solution are agitated for a time, varying in accordance with the character of the ore, generally ranging from six to twelve hours. I have extracted from complex ores, in some instances, upwards of 90 per cent of the assay-value in less than two hours, and in other instances I have found it necessary to continue the operation for twenty-four hours. No general rule can be given; each case has to be investigated and the *modus operandi* to be selected according to circumstances. (See table showing rate of extraction in relation to time of agitation, attached to the description of the Utica plant, page 94.) The strength of the cyanide solution and the volume required depend entirely on the character of the ore; as a rule, solutions for agitation should be stronger than those for percolation. Here, like in other matters in connection with cyanide treatment, experimental investigation has to advise on best conditions. (See chapter on laboratory work, page 44.) In using barrels as agitators, ore and solution will be charged before the barrel is revolved; if vertical vessels are used, the solution will be charged first, then the stirrer will be set in motion, and the ore added by degrees.

When the extraction is completed, the mass in the agitator is discharged, and the cyanide solution, now containing the gold, is separated from the solid material—i. e., the residues—by any method which local conditions and the character of the ore suggest. Apparatus of different principles have been used for this purpose. Filter presses of various constructions, vacuum filters, and centrifugal machines have been employed. Concentrates, coarse and slimy, can be successfully treated by means of my vacuum filter (see description of Sylvia and Utica plants, p. 79 and p. 89), which permits a quick filtration and a perfect and speedy washing of the residues with a minimum of liquid. In some instances I made successful use of centrifugal force for separating the gold solution and washing the residues. For washing, weak cyanide solutions from previous operations are used, and finally a water-wash is given. The residues are then discharged. The gold solutions should, for practical reasons, be kept separate according to their strength in gold and cyanide; they pass through such appliances as are used for precipitating the gold and silver, after which the "liquors" are collected in sumps for use on subsequent charges of ore. In well-appointed works no cyanide solution is allowed to run to waste, as the same amount of liquid remains constantly in circulation.

The author took out early in 1893 a caveat in New Zealand for a centrifugal apparatus, agitator and separator combined, for the treatment of slimy ores by agitation with cyanide, and subsequent separation of the gold solution by centrifugal force in the same apparatus. Experiments have of late been made in the Thames School of Mines, New Zealand, with the treatment of slimy ores by the agitation process in an apparatus which is described as follows: The appliances used in the operation consist of a shallow circular vat, a vacuum cylinder, and an air pump. The vat is provided with four revolving arms, to which soft rubber brushes are fixed. The bottom of the vat is fitted with a false bottom, constructed of a wooden grating covered with wool packing. The operation is conducted as follows: The leaching solution, made up to the required strength, is first conducted into the vat. The revolving arms are then set in motion, and the dry pulp or slimes introduced. The agitation is continued for six hours, or until the extraction is complete. A stopcock in a pipe connecting the false bottom of the leaching vat and vacuum cylinder, is then opened and the air pump started. The effect is immediate. At once the clear solution begins to drain over into the cylinder, the revolution of the arms preventing the slimes from settling and choking up the filter cloth. When the slimes have been drained down to a thick paste, the first wash water is added, the pump again started, and the slimes drained as before. The water-washings are carried on in the same way, and when completed a plug or door is opened and the leached slimes are sluiced out. The whole operation of leaching takes from eighteen to twenty-four hours. The technical and economical practicability of this method of treating slimy ores appears doubtful and will have to be proved.

B. Percolation Process.

Percolation is the method generally in use. It is being worked in the United States, in the British Colonies of Australasia, and on a very extensive scale in the South African gold fields, and therefore merits a full description. Percolation consists in soaking cyanide solution through ore. The character of the material to be treated, whether ores, concentrates, or tailings, will demand certain modifications of the treatment, without interfering with the principle.

(a) **Percolation of Ores.**—It is advisable to dry-crush the ores; the less dust produced the better for percolation. Screens of thirty meshes to the lineal inch will be found satisfactory in most instances; in some cases a coarser screen may do, or a finer one may be required. It is desirable to crush as coarse as possible without interfering with the percentage of extraction. Stamps are much in use for dry-crushing; mortars with double discharge will give more product and less dust. Rolls are to be preferred, on the ground of their giving a product of greater uniformity. The ore is charged, either directly or through hoppers, into the vats in which the percolation is conducted. These vats, or tanks, may be constructed of wood, brick and cement, concrete, iron, or steel, and vary in size in accordance with local circumstances and requirements. The largest in existence are the circular brick vats at the Langlaagate Estate and Block B Company's works in Johannesburg; these vats have a capacity of about 400 tons, and are 40 ft. in diameter by 10 ft. deep. A size in common use in Johannesburg is 20 ft. in diameter and about 6½ ft. in depth, inside measurement, of which I give Mr. J. S. MacArthur's description: "This vat is made of the best white pine; the staves are 7 ft. 3 in. long, 4 in. wide, and 2½ in. thick, and fitted with a slight taper upwards, so that the diameter at the top is about 4 in. less than it is at the bottom. The bottom is made of the same kind of material, but is at least 3 in. thick. The pieces are fitted together with dowel pins; it is then fitted into a groove cut in the staves about 6 in. above the ground, and the whole vat is bound together by steel hoops with a 6 in. over-lap and at least three rivets. No white lead or packing of any kind should be used in making these vats. If the faces of the wood are true no amount of white lead or other packing can make them truer; if they are not true, neither white lead nor any kind of packing will secure tightness. Besides this the cyanide solution being alkaline would quickly combine with and remove the oil of the white lead, if such were used, and make the vat positively worse than if none had been employed. Circular pieces, cut out of the solid wood and not bent by steam or moisture, are fixed by screws on the bottom of the vat, about 1 in. from the staves, all around the circumference. These circular pieces are about 3 in. thick and 3 in. wide, but the length of each is immaterial, provided always that a complete ring is formed. Wooden slats, about 1 in. thick and 3 in. high, are fixed about 6 in. apart all over the bottom; and an iron pipe, generally 2 in. in diameter, is screwed in from the under side near the center point. The 3 in. space from the bottom to the top of the slats is filled in with round and clean pebbles. Over this surface, formed by slats and pebbles alternately, is stretched

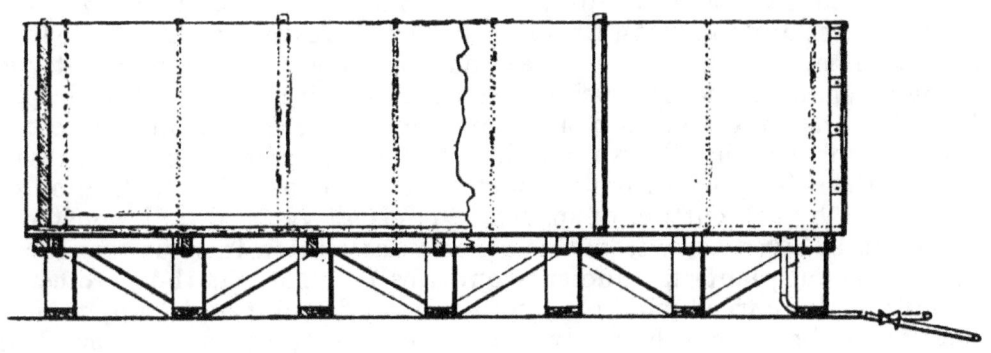

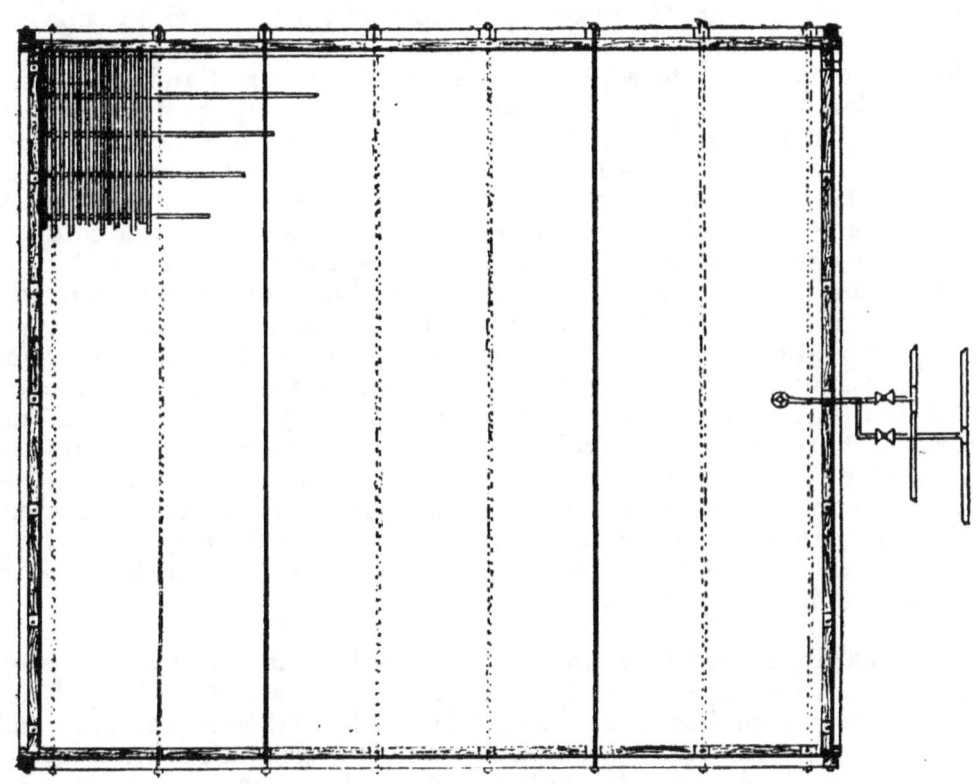

DETAILS of the FALSE BOTTOMS
— OF THE —
— PERCOLATION VATS —
— SCALE 1/4 INCH = 1 FOOT —
COP FROM W. R. FELDTMANN'S NOTES
on
GOLD EXTRACTION ETC

a canvas cloth to act as a filter, which is fastened by stretching it over a circle piece and ramming the cloth tight by pressing an inch rope into the space between the circle pieces and the staves. The canvas filter is made by shaping and sewing the canvas into a circle piece rather larger than the area of the vat bottom. In practice the canvas filter is often protected by covering it with old sacks or cocoa matting, which serves to protect the filter proper from the wear and tear caused by the friction of the ore or by the cutting of spades." (Special vat and tank constructions will be given aside from this general description in reviewing large and successful plants in different countries.) "The vat thus protected and fitted is charged with ore, and the cyanide solution is run, preferably from the bottom, by a pipe and rises slowly through the crushed ore. It must not be allowed to rush in or rise violently, as by so doing channels will be formed through which the solution will pass without acting on the ore. Such channels are apt to be formed under any circumstances, and should always be guarded against. After the upward percolation, the stopcocks are shut, and opened again after the desired time of contact has passed, so as to allow of a reversal and downward percolation. The cyanide solution now containing gold is carried through the precipitating appliances and from there into the sump, from which it may again be used for percolation. It is not wise to attempt to make the solution very rich in gold, and it is considered better practice to remove the gold frequently, as it is found that a cyanide solution containing gold is not so active as a similar solution without any."

In some instances, however, it has been found of advantage to use gold cyanide solution over again, without first passing the same through the precipitation boxes. (See experiments made by the Robinson Company, in Johannesburg, page 65.) According to the richness of the ore and the fineness of the grinding, percolation may be repeated several times, but after the final percolation with the ordinary cyanide solution, a washing of weak or waste solution should follow, and the whole operation be completed by a water-wash, after which the residues are discharged. The filling and discharging can be done either by hand or by mechanical appliances. The various methods will be described in connection with the process in South Africa.

(b) **Tailings** are treated in substantially the same way as ores, and, the quantities being large and the grade low, the vats of the largest size and the most complete arrangements for saving labor in charging and discharging are necessary for profitable working; generally speaking, the difficulty of discharging the vats is increased by the increase of their diameter and their depth. "Several difficulties arise in the case of tailings, which do not usually present themselves with ores. These difficulties are chemical and mechanical. The chemical difficulties have been described in the chapter on chemistry; no general rules can be applied to them; each case has to be investigated and steps be taken accordingly. The mechanical difficulties arise from the tailings being derived from the operation of wet-crushing. When tailings are charged in a wet state into the percolating vat, they are apt to remain in lumps, from which the water has to be expelled by the cyanide solution before the latter can effectually do its work. It is obvious that where tailings are already saturated with water, the cyanide solution will have a difficulty in penetrating, and this difficulty is increased when the wet tailings are held together in masses between which the cyanide finds an easier channel for flowing than by soaking through them. This is merely a form of the channels above referred to. Assuming, however, that these channels are not formed, the tailings in a wet state mass or pack together to such an extent as seriously to retard percolation. Another difficulty which arises from the tailings being wet is that in clayey ores the slimy portion of the mass is apt to gather into a layer by itself, which if formed of real clay, not only impedes but absolutely prevents percolation. In order to overcome this difficulty the simple method of drying and mixing should be adopted. The drying is a mere preliminary to the essential of thorough mixing. Particles of clay, which are not kept apart by sand, will agglomerate and form water-proof strata. It is impossible, unless the whole material is perfectly dry, to get the particles of clay separated from each other and allow the sand particles to intervene. Even when this is done, the tendency of the clay particles to agglomerate must be guarded against and prevented. The principal precaution necessary is that the solution, whether applied from top or bottom, should not flow more quickly than the dry tailings can absorb it. In many respects upward percolation has an advantage, but principally because the flow of solution is against gravitation. In downward percolation, where the flow of solution and gravitation act together, the whole material tends to become compressed into a cement, through which the solution penetrates but slowly, preferring to take the easier course down the sides of the vat, and in fact going around rather than through the tailings. Alternate upward and downward percolation may be found useful in some cases."

The percolation vats are charged with tailings to within a few inches from the top, and their surface is leveled. The cyanide solution of, say from 0.2 to 0.8 per cent of strength, is then permitted to penetrate the tailings, till the liquid covers them. The contents of the vat will settle some inches, which shrinkage depends on the depth of the vat and the percentage of moisture in the material. The solution is permitted to remain undisturbed in contact for say twelve hours; after that time it is allowed to drain off. As the liquid is drawn off, it is replaced by fresh solution. This operation is continued for a longer or shorter period in accordance with the value of the tailings (about six to twelve hours in the works of the Robinson Company at Johannesburg). After this time, which is termed the "strong solution leaching," a weaker solution, containing say from 0.2 to 0.4 per cent of cyanide, is turned on, which filters through the ore for about eight to ten hours. This weak solution, when drawn off, is treated separately (see above). At last, water is run on the tailings for replacing the last weak cyanide solution. The volume of solution in constant use and circulation remains the same. The weak cyanide solution is the liquor which has previously passed through the process by which the gold and silver are precipitated and has from the sumps been pumped back into the vat.

The percolation vats, which used to be square, are, in new works, round. The cyanide solution of the described strength has no appreciable deleterious effect either on the wood of the vats or on the iron pipes and iron valves of the pumps. Iron or steel vats may be protected by a coating of coal tar and asphalt, or a solution of asphalt in turpentine, preferably put on hot, if special reasons make such protection desirable. The quantity of cyanide solution used for the treatment of

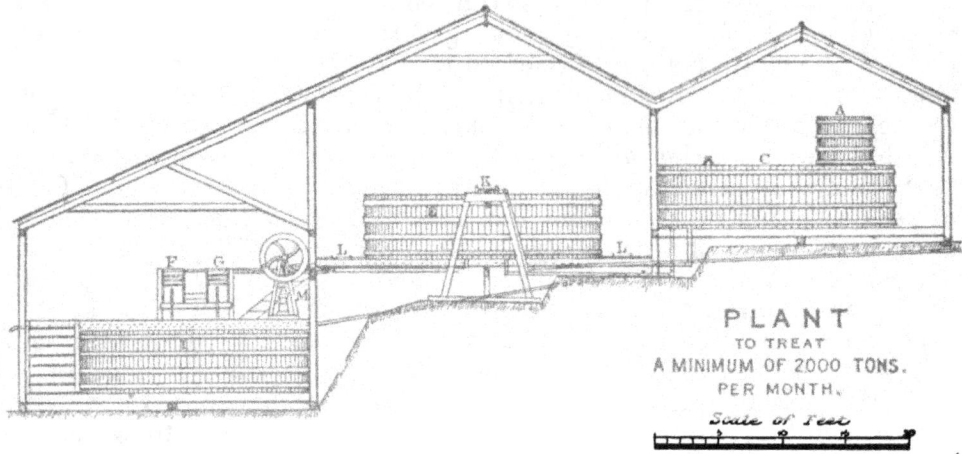

PLANT
TO TREAT
A MINIMUM OF 2000 TONS.
PER MONTH.

Scale of Feet

one ton of tailings amounts generally to half a ton of strong solution and half a ton of weak solution and wash.

When the percolating process is finally completed, the exhausted tailings, or "residues," are discharged in older works by being shoveled out over the side. More modern works have trap-doors at the bottom of their vats for discharge (see diagram, page 59). "Sluicing out" of the residues is being practiced in several localities. The large vats of the Langlaagte Estate Company's works at Johannesburg, which hold 400 tons of tailings each, are discharged by means of running cranes (see diagram, page 57).

It must be always borne in mind that the most complete arrangements for saving labor in charging and discharging large quantities of low-grade material are necessary for profitable working. In order to achieve that end, a plant should offer all such facilities which circumstances permit; it should be so arranged that the tailings will not have to be lifted, but can be dumped into the percolation vats. The size of the vats has been constantly increased. New works, like the Roodeport works at Johannesburg, are supplied with vats 40 ft. in diameter. The table appended illustrates the dimensions of the percolation vats in some of the more important cyanide extraction plants:

Sizes and Material of Percolation Vats.

Name of Works.	Location.	Number of Vats.	Material.	Form.	Dimensions.	Contents in Cubic Feet.	Contents in Tons.	Class of Material Treated.	Time of Operation in Days.
1. Crown Mines Company	New Zealand	24	Wood	Square	11 ft. by 9 ft. by 3.75 ft.	371.25	7	Ore	---
2. Great Mercur Company	New Zealand	3	Wood	Square	16 ft. by 12.5 ft. by 3.5 ft.	700	20	Tailings	---
3. Tryfluke Company	New Zealand	4	Wood	Square	12 ft. by 16 ft. by 4 ft.	768	20	Tailings	---
4. Waihi Company	New Zealand	13	Wood	Round	Diam. 22.5 ft., 4 ft. deep.	1,589.6	30	Ore	7
5. Cripple Creek Gold E. Company	Colorado	4	Iron	Round	20 feet diam.	---	---	Tailings	---
6. Revenue Company	Montana	---	Wood	Round	10 ft. diam., 4.5 ft. deep.	353.26	---	Tailings	1 to 1.5
7. Mitchell Creek G. M. Company	N. South Wales	6	Wood	Round	18 ft. diam., 5 ft. deep.	1,271	35	Tailings	3
8. African Gold Recovery Comp'y	Transvaal	---	Wood	Square	---	---	35 to 50	Tailings	---
9. Crown Reef Company	Transvaal	---	Wood	Square	16 ft. by 18 ft. by 6 ft.	1,728	---	Tailings	3 to 3.5
10. Crown Reef Company	Transvaal	---	Brick	Square	40 ft. by 40 ft. by 10 ft.	16,000	---	Tailings	---
11. Durban-Roodeport Company	Transvaal	---	Wood	Round	40 ft. diam., 7 ft. deep.	8,792	---	Tailings	---
12. Nigel Company	Transvaal	---	Wood	Round	16 ft. diam., 5 ft. deep.	1,004	---	Tailings	---
13. Nigel Company	Transvaal	---	Wood	Square	16 ft. by 24 ft. by 5 ft.	1,920	---	Tailings	---
14. Langlaagte Estate Company	Transvaal	---	Brick	Round	40 ft. diam., 14 ft. deep.	17,584	400	Tailings	---
15. Robinson Company	Transvaal	12	Wood	Round	---	2,000	75	Tailings	---
16. Simmer & Jack Company	Transvaal	---	Wood	Round	42 ft. diam., 14 ft. deep.	23,386	---	Tailings	---

All vats should be some distance above the ground, so that leaking can be easily detected; concrete foundations for the vats are generally adopted. The wooden tank material is an absorbent of both cyanide and gold, particularly when new. It has been found at the Salisbury works, Johannesburg, that pine wood lying thirty-four hours in a 0.3 per cent cyanide solution reduced it to 0.05 per cent, while cement reduced it to 0.24 per cent. Cement tanks have come into use of late, and have proved satisfactory; such tanks and vats may advantageously be built into excavations in solid ground. Many attempts have been made to discharge tailings-pulp direct from the plates, or ore-pulp direct from the mortars into the percolating vats, but their successful treatment by cyanide when so discharged has been prevented by mechanical causes, the reason being that the material packs so densely in the vats that it makes percolation an extremely tedious operation, and in consequence of the presence of slimes the results are unsatisfactory. The advantages of wet-crushing over dry-crushing are, from an economical standpoint, so obvious, however, that experiments will be continued, and ultimately the drawbacks which now adhere to the method may be overcome. Cyanide of potassium solution has been used, in some instances and in an experimental way, in lieu of water in the mortars, when wet-crushing was resorted to, but does not appear to be practiced anywhere at present. An innovation in percolation consists in the circulating system, which will be described in detail in connection with the practice of the cyanide process in South Africa (see page 46).

(c) **Percolation of Concentrates.**—"These are treated similarly to ores, but being generally richer require a greater number of percolations, and thereby a much longer time. In most cases, their quantity is limited, and the size of the percolation vats varies in accordance with the quantity." I have, in most cases, given agitation the preference to percolation for treatment of concentrates, on account of its greater cheapness and rapidity; in Africa most companies prefer the latter method. Percolation of concentrates requires about twenty days, the reason for which will be found partly in the coarser character of the gold, partly in its being in the form of amalgam, and mainly in the difficulty the solution has in penetrating between the faces of the sulphuret crystals. A difficulty sometimes arises in the percolation of concentrates, owing to the crystalline form of iron pyrites and galena. These minerals crystallize in cubes, and when suspended partially or wholly in a fluid tend to range themselves face to face, so that a section of such a mass deposited from a fluid would resemble a brick wall in structure. This difficulty does not arise in the case of sand or minerals which crystallize in other systems. Whenever it occurs, it may be overcome by mixing the cubical sulphurets with coarse sand.

C. Cyanide and Cyanide Solutions.

The best strength of solutions to use in either percolation or agitation depends entirely upon the nature of the ore, and it is impossible to set any rule. The strength of solutions generally used varies from one eighth to one per cent of cyanide. (In reference to the determination of the correct strength to be used in treating any class of ore, see chapter on laboratory work, page 44.) "For convenience and economy of work, the solutions are generally divided into three classes: No. 1, No. 2, and No. 3, of which No. 1 is the strongest and No. 3 the weakest. Assuming that the material under treatment does not require a preliminary alkaline wash, or that such treatment has already been completed, it is usual to run on a weaker solution, say No. 2, in the first place, and after its percolation to use No. 1, and then No. 3 in the same manner, finishing with a water-wash, the first portion of which is run into and forms part of the No. 3 solution. These different solutions are kept separate after percolation, and when charged with gold are subjected by themselves to the precipitating process. In some works sumps are used as reservoirs, and the solution is pumped direct from them onto the ore; but space permitting, it is considered better practice to have reservoirs for each solution above the percolation vats, from which the flow can be more easily regulated." For the purpose of bringing weak solutions up to a certain standard, it is advisable to use a very strong solution, of which enough is added to bring the weak solution up to the required strength. This method has to a great extent taken the place of the old method of using solid cyanide to bring weak solutions up to high standards. The strength of the cyanide solutions, which it is of great economical importance to determine, is tested according to Liebig's method by means of a one tenth standard solution of nitrate of silver, which is made by dissolving seventeen grams of pure nitrate of silver in one litre (1,000 cc.) of distilled water. Liebig's method is based on the fact that silver cyanide is soluble in excess of potassium cyanide, with formation of a double cyanide of silver and potassium:

$$KCy + AgNO_3 = AgCy + KNO_3$$
(Potassic cyanide.) (Argentic nitrate.) (Argentic cyanide.) (Potassic nitrate.)

$$AgCy + KCy = KAgCy_2$$
(Argentic cyanide.) (Potassic cyanide.) (Argentic potassic cyanide.)

As soon as the whole of the cyanide has been converted into a soluble silver salt, an additional drop of silver nitrate will produce a permanent precipitate of the insoluble simple cyanide of silver:

$$KAgCy_2 + AgNO_3 = KNO_3 + AgCy$$
(Argentic-potassic cyanide.) (Argentic nitrate.) (Potassic nitrate.) (Argentic cyanide.)

A measured portion of the perfectly clear cyanide solution which is to be tested is taken; if necessary some distilled water is added, and the standard silver solution is gradually added from a graduated burette, until a permanent white cloud is formed. As each cubic centimetre of the silver solution is equal to 0.013 grams of potassium cyanide, by multiplying the number of cubic centimetres consumed by 0.013 the amount of cyanide in the solution tested is found in grams, from which the percentage can easily be calculated. A convenient silver solution for the purpose of analyzing cyanide solutions is one of such strength that every cc., added to 10 cc. of the solution which is to be tested, corresponds with 1 per cent pure cyanide of potassium. "The cyanide solutions are apt to form, by continued exposure to the air, carbonate of ammonia; and as this salt interferes very seriously with the determination of the cyanide, it is well to add a few drops of solution of iodide of potassium, which forms a pale-yellow cloud insoluble in ammonia,

which indicates completion of the reaction" (MacArthur). The mode
of analysis, as described above, calculates the amount of cyanide of
potassium in a solution by ascertaining its contents of cyanogen. If the
cyanogen is partly combined with sodium instead of potassium, the percentage of cyanide appears in the analysis higher than it really is; the
value of commercial cyanide of potassium should therefore be ascertained by determining its contents of sodium, if any, as it is possible,
by manufacturing a mixture of cyanide of potassium and cyanide of
sodium, to produce cyanide, which according to the ordinary method of
estimation, contains apparently more than 100 per cent of potassium
cyanide. The analysis of cyanide solutions for gold and silver will be
described in the chapter on laboratory work (page 44).

The Cyanide usually employed for ore extraction is of two classes, one
of which, manufactured in Scotland, contains from 70 to 80 per cent of
pure potassic cyanide; the other is manufactured in Germany and contains upwards of 98 per cent. The latter grade is preferable, because it
contains no carbide of iron, the presence of which in the former not only
involves periodical cleaning out of the dissolving tank, but also is liable
to precipitate gold should any of it come in contact with gold cyanide
solution. The price of the best quality of cyanide of potassium, guar-
anteed to contain upwards of 98 per cent cyanide, is at present 50 cents
per pound in the United States, delivered at seaports.

D. Treatment of the Gold Solutions (Recovery of the Bullion).

All methods of treating ores, as described, yield solutions containing
more or less gold and more or less cyanide of potassium. The next
step is to recover the gold and silver. The chief method for that end
consists in their precipitation by means of finely divided zinc; it forms
one of the patent-claims of MacArthur and Forrest, and is generally in
use. Another method is that of B. C. Molloy, of Johannesburg, who
decomposes the cyanide solution of gold by means of an alkali metal,
and amalgamates the bullion thus liberated; cyanide of potassium is
regenerated by this process. The Molloy process has been in use on a
small scale in South Africa, but has apparently gone into disuse again,
for the official list of the Chamber of Mines of Johannesburg for March
does not return any gold as extracted by that process, as had been the
case in former months. Other processes for bullion precipitation are:
The Siemens and Halske process of precipitating the noble metals by
electrolysis on lead sheets, which method is in use at the Worcester works
in Johannesburg; the Pielsticker process of using electrolysis, constantly
applied to the circulating solution; the Moldenhauer process of precipitation by means of aluminium, and the Johnston process of using charcoal as reducing agent. No information is available on the technical
application of the three processes last named.

(a) Bullion Precipitation by Zinc.—The solutions from the percolation vats, or from the filter appliances if the agitation process is used,
are run through boxes which are divided into chambers with double
partitions. The first partition does not reach to the bottom of the box,
the next one not quite to the top, and so on; they compel the solution to
enter each chamber from below, and pass through a perforated bottom,

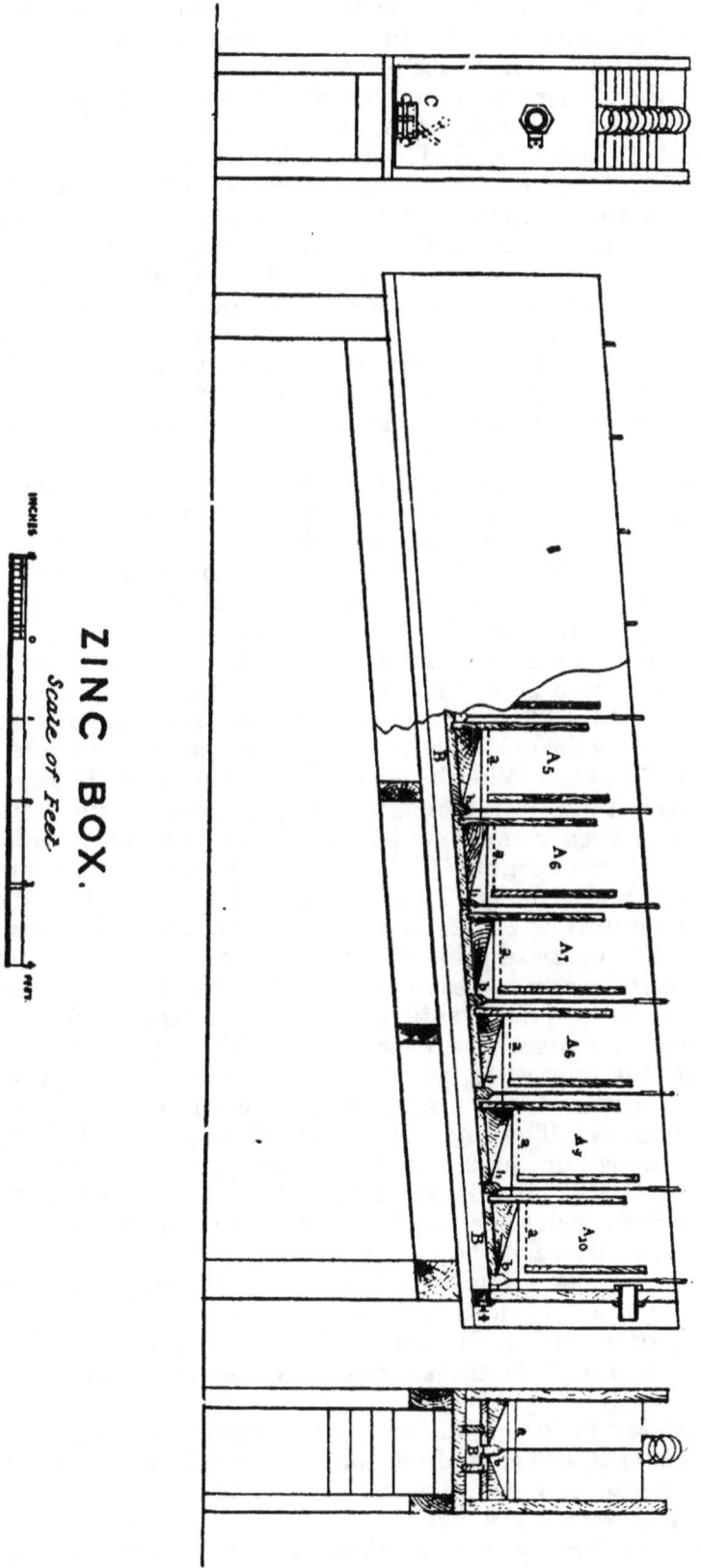

ZINC BOX.

Scale of Feet.

on which the finely divided zinc is placed. After passing through the
zinc, the solution leaves the chamber at its top; then it descends between
the double partition to the space below the perforated bottom of the succeeding chamber, where it undergoes the same treatment, and so forth
in all successive ones (see sketches, pp. 31, 33). "This arrangement has
been adopted because the gold is precipitated on the zinc in a state of
fine division, and would, if deposited on the upper surface, prevent the
further flow of the solution; but being deposited on the under surface,
the gold precipitate falls off and leaves the passage clear. Each precipitating box may contain ten to twelve double chambers, and no matter
how rich the solution is at the inflow, it should not contain more than
a few grains per ton at the outflow." In some works the zinc boxes are
up to 40 ft. long. There is, however, no advantage in going beyond a
limited number of chambers, as precipitation of the metals takes place
chiefly in the first few compartments of the box. (See description of Utica
plant, p. 89).

I give here Mr. J. S. MacArthur's description of his construction of
filter boxes, and the mode of working them which is used by the
MacArthur-Forrest patentees in South Africa: "The gold precipitate falls
through the gauze 'a' (see cut, p. 31) into a chamber which communi-
cates with an inclosed launder or gutter. From day to day fresh zinc is
added, always adding it in the last chamber and bringing the partly
consumed zinc up a step, so that the first chamber contains zinc half
consumed and rich in gold, while the last chamber contains fresh zinc
containing no gold. At intervals of about two weeks, there is a clean-up,
and the gold is collected by stirring the zinc so as to cause the gold
precipitate to fall off. When this is done, the stopper 'b' is raised and
the gold precipitates fall through the opening into the launder 'B.'
When this has been completed for each chamber, the launder is discharged through the opening 'C.'" The precipitation boxes are usually
made of wood, and although I have been well satisfied with such material (kauri-pine in New Zealand), I substituted it in California by steel,
which I found in every respect an excellent material for the purpose. I
could not ascertain any increased loss of cyanide by the use of iron as
box material; the galvanic action of iron and zinc in contact on the
cyanide seems by some writers overrated (see description of Utica plant,
p. 89). Of this apparatus, which is based on the same principle, but
which in its construction is simpler than the one described, I give the
appended diagram. The total length of the apparatus is 9 ft., the size
of the chambers is 9 in. by 9 in. by 14 in. deep; the distance of the partitions between each chamber is 1 in. The perforated and movable
false bottom of each chamber is of steel, which is an advantage over wire
sieve bottoms, which easily become clogged by bullion. The real bottom
of each chamber has a faucet of 1 in. diameter, which discharges the
liquid and the finely divided bullion into a tank below the apparatus,
from where, after settling (under addition of some alum, if saving of
time is an object), it is transferred to a vacuum filter, as described
further on.

"The zinc used for precipitation purposes should be the best quality
found in commerce, and should not contain arsenic or antimony; a small
percentage of lead, however, does no harm, but rather tends to promote
rapid action by forming a voltaic couple with the zinc" (MacArthur).
The metal is preferably used as shavings, or filiform, as these forms

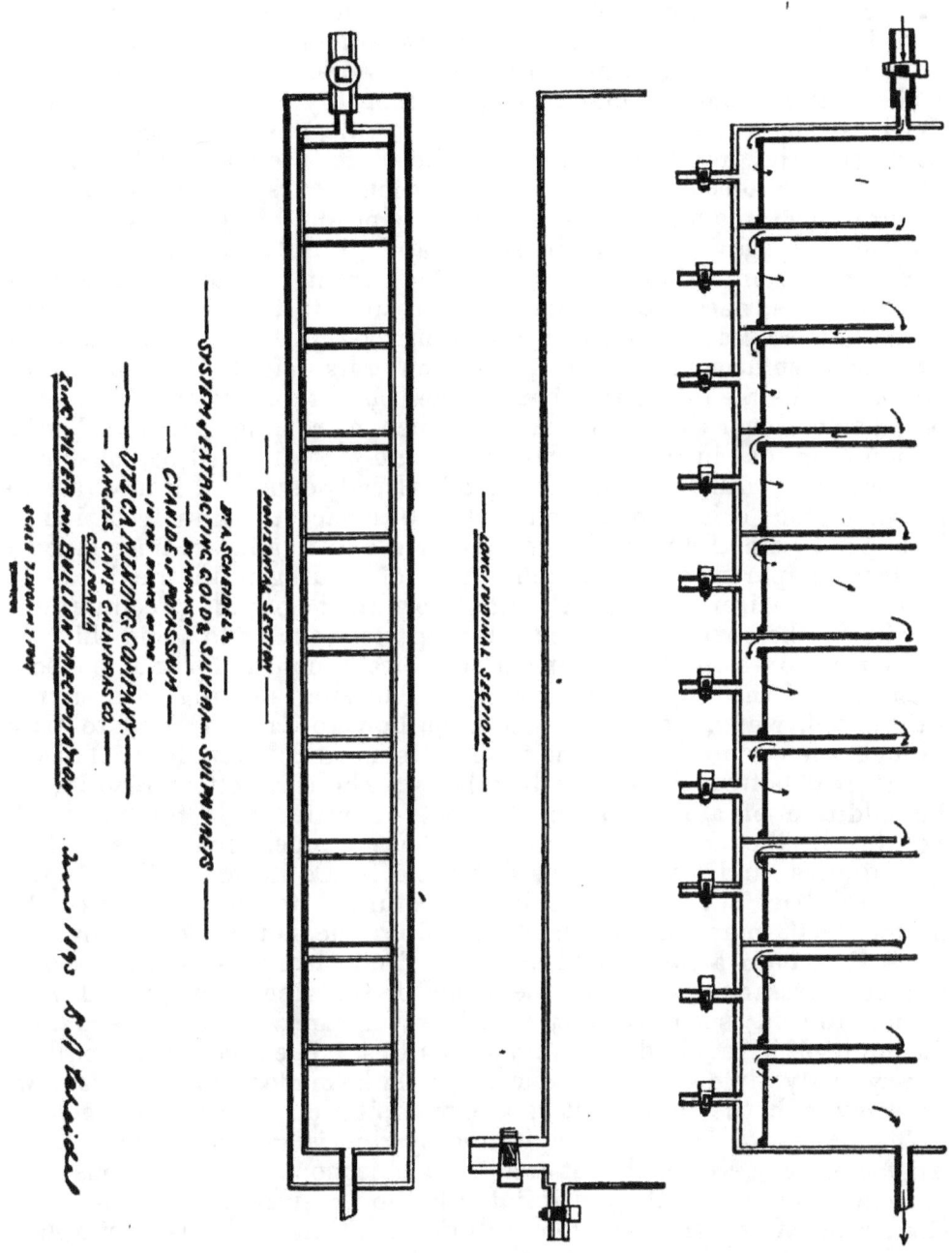

give in practice the most surface for the least weight and do not pass readily through a sieve, whereas the gold, which is precipitated as a fine powder, does. Shavings have the advantage of not forming lumps so easily in the precipitation boxes as filiform zinc; the latter has, however, the greater advantage of being cheaper in its preparation, as no remelting of the commercial zinc is required. It is prepared by cutting sheet zinc into disks, a number of which are placed together and turned on a lathe with an ordinary chisel. The zinc linings of the cyanide packing-cases, for which there is no market, may be turned to account in that way.

3CP

In reference to the cost of preparing the zinc, I may quote the Nigel Gold Mining Company in Johannesburg, where one native, working about eight hours a day, can easily keep the works going, with an output of about two thousand ounces of gold monthly; the consumption of zinc is about twenty pounds daily. As a rule, one cubic foot of zinc shavings in the precipitating box is sufficient for the precipitation of the gold from two tons of solution per twenty-four hours, or, roughly speaking, from the same weight of ore (see the zinc for bullion precipitation in Africa on page 53). Zinc in sheets and granulated zinc have been tried for bullion reduction, but with indifferent results, on account of their limited surface. Zinc amalgam and zinc dust have not answered, for mechanical reasons—zinc dust packing too tightly and zinc amalgam not offering sufficient surface in proportion to its weight. The precipitation of the metals in the zinc boxes takes place rapidly; the zinc in the compartment near the influx will be much more quickly charged with bullion than that in those more distant, and the zinc will be consumed in proportion. Zinc on which bullion is already deposited is more active than new zinc; it is therefore advisable to replace the dissolved zinc in the upper by zinc from the lower chambers, and to add the fresh zinc in the last compartment. The generation of hydrogen in the boxes is liable, by polarization, to partly interfere with the bullion precipitation; the zinc in the boxes should be stirred up occasionally to avoid this.

The zinc boxes are cleaned up once or twice a month; for that purpose the inflow of the solution is stopped. The zinc shavings are stirred with a rod, which causes the fine bullion to fall off and to pass through the perforations of the false bottoms, and through the faucets at the real bottoms, into the box below, where it settles readily, on the addition of a little alum. A jet of water will further wash the zinc in the chambers. This method of operating takes only a few minutes, and has been used by me in California. The liquid standing above the settled bullion is returned to the zinc box; the bullion itself, unavoidably mixed with fine zinc, is transferred through a fine sieve onto a vacuum filter. If a final cleaning-up is desired, it will be necessary to dissolve the whole of the zinc, impregnated with bullion, in acid; such necessity will, however, rarely arise. The manipulation itself if required, offers no difficulties. The precipitated bullion is very finely divided, and provision should be made to prevent its flowing away with the liquid out of the precipitation boxes (see page 52).

The process of bullion precipitation by zinc is, generally speaking, a satisfactory one, although not free from objections; all operations with and manipulations of the precipitated bullion require care to avoid loss. The action of the zinc on gold solution is theoretically very simple— a simple substitution of gold by zinc according to the equation:

$$2 \text{ KAuCy}_2 + \text{Zn} = \text{K}_2\text{ZnCy}_4 + 2 \text{ Au}$$
(Auro-potassic cyanide.) (Zinc.) (Zinc-potassic cyanide.) (Gold.)

One pound of zinc should precipitate about six pounds of gold. The actual consumption is, however, considerably larger, and amounts to from 5 oz. to 1 lb. of zinc per ounce of gold recovered. A constant generation of hydrogen gas in the precipitation boxes proves the effect of the potassic hydroxide on the zinc, and probably a decomposition of cyanide of potassium, going on parallel with the decomposition of the double gold cyanide. A considerable loss of zinc occurs generally in refining the precipitated bullion, which always contains a high percentage of that metal. The double salt of auro-potassic cyanide appears to be one of the most stable of gold salts; its decomposition by zinc is, however, practically complete; an excess of cyanide of potassium in the solution does not redissolve precipitated gold in the boxes as long as there is zinc present. The cyanide of potassium formed into a double salt with zinc during the gold-reducing process is not available for dissolving gold in new operations. If a surplus of caustic soda has been used for neutralizing acid salts in the ore without following washing, the loss of zinc will naturally be increased. A white precipitate is constantly accompanying the reduction of bullion in the zinc boxes, undoubtedly the result of the action of alkali on the zinc and of the zinc-potassium oxide on the double cyanide of zinc and potassium, which is always present in the solution, forming the insoluble cyanide of zinc; ferrocyanide of zinc is also formed in the boxes. Ferrocyanide of zinc is formed in the percolation vats when the double cyanide of zinc and potassium comes in contact with the iron salts in the ore, and, as it is insoluble, to this cause is due the constant removal of zinc from the solution with the residues (Buckland). The gold precipitate on the zinc is, as a rule, brown to black, with sometimes a metallic luster; it is mostly slimy, and when dry it seldom contains more than 40 or 50 per cent of gold and silver, the remainder being finely divided zinc and its accompanying impurities, such as carbonate of lead. It may also contain copper, if that metal is present in solution. (In the instance of treating concentrates containing carbonate of copper from the Sylvia Mine, New Zealand, the gold solution contained a very appreciable quantity of copper; this complicated matters by causing the copper to precipitate with the gold and cover the zinc, thus forming a galvanoplastic coating, which made it necessary to dissolve the whole of the zinc for the purpose of obtaining the bullion, till I found an addition of cyanide to the solution, before it enters the zinc boxes, useful for the prevention of the deposition of the copper.)

I always found mercury in the zinc bullion in not inconsiderable quantities when concentrates had been treated by agitation; such mercury must have been derived from amalgam, and mercury saved with the pyrites on the concentrators. Gmelin and others describe mercury as absolutely insoluble in cyanide. There is, however, no doubt of the correctness of my observations; the mercury must have been dissolved in the cyanide solution, which entered perfectly clear into the zinc boxes; solubility of gold amalgam in cyanide may offer the explanation. Traces of antimony and arsenic have also been found in the bullion. The precipitation of silver goes on by zinc simultaneously with the gold; it is even more rapid and complete than that of gold. (See my table on the process in the Utica Works, page 91.)

Other apparatus than the described boxes have been suggested for bullion precipitation by zinc—for instance, earthenware and porcelain vessels have been recommended. They have the apparent advantage of cleanliness; their construction, however, makes the cleaning-up of the bullion difficult, the connection between the single cells being complicated; they have not become a practical success in works of any extent; they are all based on the principle of the solution penetrating the zinc from below and running off at the top. The precipitates, obtained as

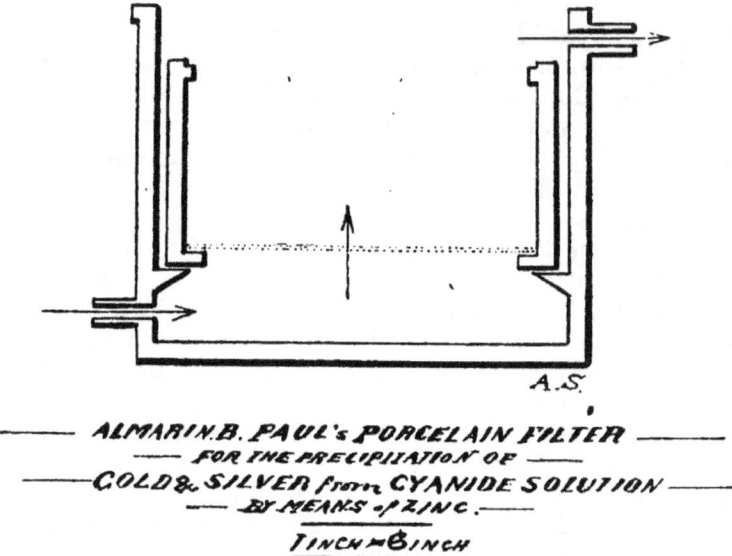

———— ALMARIN B. PAUL's PORCELAIN FILTER ————
—— FOR THE PRECIPITATION OF ——
——GOLD & SILVER from CYANIDE SOLUTION——
—— BY MEANS of ZINC.——
1 INCH = 6 INCH

described from the zinc boxes, are transferred to a sieve, made of No. 1 punched battery screen or a 40-mesh wire screen, through which they are washed onto a filter in connection with a vacuum chamber, where they are liberated from the adhering cyanide solution and reduced from their very voluminous state into a more compact form. This filtration will always be found slow on account of the extremely slimy character of the bullion. For filtering and washing the bullion slimes, filter presses may be suggested. By the screening process the coarser particles of zinc are separated from the bullion, but the bullion still contains a large percentage of very fine zinc, of which it is advisable to remove as much as possible before melting.

Bullion Refining.—The means for this purpose are calcination or roasting and acid treatment. I use for roasting (see description of Sylvia plant, p. 79) a muffle furnace, where the slimes are dried, and then calcined for the promotion of the oxidation of the base metals. The calcining process is generally in use in South Africa, and will be described with the cyanide practice in Johannesburg, page 54. I generally prefer sulphuric acid treatment, with following washing and drying of the bullion. The acid treatment is a comparatively simple operation, and does not require, even for large quantities of bullion, any other apparatus than wooden tubs, the increased temperature produced by the reaction of the acid on the zinc making application of artificial heat superfluous. The separation of the acid solution from, and the washing of, the bullion is best done by decantation, and completed on the bullion filter mentioned above. It is advisable to liberate the bullion as much as possible from base metal before melting, which is otherwise connected with loss of gold by evaporation caused by the volatilization of zinc; besides, the zinc fumes are very disagreeable. The presence of oxides of base metals (as obtained by calcination) makes the melting tedious and expensive on account of the detrimental influence of the slag on the melting pots. The presence of a high percentage of base oxides prevents the use of graphite crucibles and compels the use of clay pots. Bullion, when treated by acid as described, does not offer any difficulties in melting, if proceeded with in the following manner:

Melting Room for Cyanide Bullion in Sylvia Company's Works, Tararu, New Zealand.

The bullion, after having been treated with sulphuric acid and washed with water, is dried by suction on the vacuum filter as much as possible, after which it is easily detached from the filter cloth. The mass is then charged into the muffle (see plans of Sylvia cyanide plant, page 77). The heat is kept low to drive off the moisture; it is then increased gradually to dark red; after about one hour's calcination, during which time the oxidation of base metals which escaped removal by acid treatment is going on, the mass presents a gray-brown appearance. Attention has to be paid to the draught to prevent loss of the fine precipitates. Bullion resulting from the treatment of concentrates will invariably give off quicksilver vapors; condensation yielded only a small quantity. The calcination process completed, the roasted bullion is carefully transferred from the muffle, by means of a small shovel, into a wrought-iron box for cooling purposes. It now presents itself in the form of lumps, approaching more or less the spherical form, of the size of peas, largely mixed with dust. When sufficiently cooled, it is charged into a pulverizing cylinder of sheet-iron, 3 ft. long and 2 ft. in diameter, which is revolved by means of a pulley. Large pebbles are charged into the apparatus with the bullion to aid pulverization. Borax, with preference borax glass, and soda ("ammonia-process soda") are added into the barrel, in proportions according to experience, for securing a fusible clear slag of light specific gravity. If the bullion is base on account of a large proportion of zinc oxide, which happens only if acid treatment was not properly conducted, a silicious flux, like sand or glass, has to be added. Acid sulphate of soda and fluorspar have been occasionally found useful as additional fluxes. During pulverization, a thorough mixture of bullion and fluxes will take place. Moisture in the fluxes should be avoided, as it is a certain source of loss in melting, the escaping water carrying fine bullion out of the pot.

Plumbago crucibles are very well adapted for melting bullion, prepared as above described; they stand almost as many operations as with battery gold. In melting, some borax is first put into the crucible; the bullion mixture from the pulverizing and mixing cylinder is not added all at once, but as each portion melts and sinks down, fresh quantities of it are put in. The melting goes on speedily and in the most satisfactory manner. When the whole quantity is finally charged, the temperature is kept very high for some time, to give the small bullion globules a chance to collect. The contents of the pot are poured into a heated mold in the usual manner. No chemical losses will be experienced by this method of bullion melting. Bullion obtained by the described manipulation will be found to be at least 950 fine. The bullion produced by cyanide works generally varies in fineness according to the attention paid to its refining. Gold purchasers buy bullion on assay, and refiners charge higher rates for baser bullion; it is therefore as a rule cheaper in the end to produce clean bullion. Bullion precipitation by means of zinc is not free from objections; its practice is connected with a loss of cyanide and with the introduction into the process of a new compound (the double zinc-potassium cyanide), which is, to say the least, not an advantage; the consumption of precipitating agent (zinc) is far in excess of the amount theoretically required, and no opportunity is offered for a regeneration of the cyanide. The precipitation itself is, however, a very efficient and simple procedure, not requiring either motive power or more than ordinary attention, and the treatment of the precipitates cannot be said to offer any obstacles which would characterize the process as metallurgically inefficient. In fact, generally speaking, the clean-up of the bullion precipitated by zinc, if properly handled on the lines explained above, will hardly be found more troublesome than a mill clean-up.

Although the precipitation of the gold by zinc is unquestionably a weak point in the cyanide process, no other method has as yet taken its place to any extent. Had other methods, which require motive power, careful adjustment of costly and delicate machinery, and constant attention preceded it, its discovery would probably have been considered an improvement of importance.

(b) The Molloy Process.—A method for precipitating bullion which has obtained some technical importance has been advanced by B. C. Molloy, of Johannesburg, whose process is protected by English letters patent No. 3,024, dated 16th February, 1893. Molloy uses for precipitation purposes sodium or potassium amalgam, which is formed electrolytically from a solution of carbonate in contact with a bath of mercury. The alkali metal combines with the cyanogen of the gold compound, forming an alkali salt of the cyanogen, while the gold is instantly amalgamated. This auriferous amalgam is then strained and melted as in an amalgamation mill. The process of precipitating and collecting the bullion is carried out in an amalgamating apparatus, the bottom of which is partially covered with mercury. On this mercury rests the solution from which the metals are to be precipitated. The mercury is charged electrolytically with an alkaline metal (by the electrolysis of an alkaline salt used in a porous vessel in contact with a mercury cathode). The alkaline metal, or its amalgam, when coming to the surface of the mercury and in contact with the water of the solution, decomposes the water, the alkaline metal combining with the oxygen of the water to form an alkaline oxide; the hydrogen of the decomposed water is at the same time evolved in a nascent state from the surface of the mercury which is in contact with the solution from which the gold is precipitated, and absorbed by the mercury. The gold is released from the mercury in the ordinary manner by straining and distillation. Another, "though much less advantageous method," suggested by Molloy, is the mechanical addition to the mercury of potassium or other alkaline metals, or amalgam of the same. In both cases the original solution of cyanide of potassium is regenerated and ready for use again. The reaction is as follows:

$$K_2CO_3 + \text{elect. current} = K_2 + CO_2 + O.$$
$$KAuCy_2 + K = Au + 2 KCy.$$

No information could be gained of the actual working results.

Other methods of bullion precipitation by means of the electric current have been suggested.

(c) The Siemens and Halske Process for bullion precipitation by electrolysis on lead sheets has become of technical importance. The precipitation plant consists in boxes through which the gold solution passes; these contain the anodes, which are iron plates, and the cathodes, which are lead sheets, stretched between iron wires fixed in a light wooden frame, which is suspended between the iron plates. It is claimed that the adoption of this method of bullion precipitation permits the use of very weak cyanide solutions for extracting purposes, followed by a considerable reduction in the cost of treatment.

Mr. A. von Gernet has given the following details as to the practical working results obtained at the Worcester works, in Johannesburg, with the Siemens and Halske process, which has been tried there for four months on a large scale, after long and exhaustive preliminary experiments at the works of the Rand Central Ore-Reduction Company:

"There are now in use five leaching vats of 2 ft. in diameter with 10 ft. staves, each holding 2,700 cub. ft. One tank is discharged and filled every day. The strong solution used contains from 0.05 to 0.08 per cent cyanide, and the weak washes 0.01 per cent. The actual extraction of fine gold has averaged 70 per cent, while the consumption of cyanide has been ¼ lb. per ton of tailings treated.

"The precipitation plant consists of four boxes 20x8x4 ft. Copper wires are fixed along the top of the sides of the boxes, and convey the current from the dynamos to the electrodes. The anodes are iron plates 7 ft. long, 3 ft. wide, and ⅛ in. thick. They stand on wooden strips placed on the bottom of the box, and are kept in vertical position by wooden strips fixed to its sides. In order to effect circulation in solutions passing through the box, some of the iron sheets rest on the bottom, while others are raised about 1 in. above the level of the solution, thus forming a series of compartments similar to those of a zinc box, the difference being that the solution passes alternately up and down through successive compartments. The sheets are covered with canvas to prevent short circuit. The lead sheets are stretched between two iron wires, fixed in a light wooden frame, which is then suspended between the iron plates. The boxes are kept locked, being opened once a month for the purpose of the 'clean-up,' which is carried out in the following manner: The frames carrying the lead cathodes are taken out one at a time. The lead is removed and replaced by a fresh sheet, and the frames returned to the box, the whole operation taking but a few minutes for each frame. By this means the ordinary working is not interrupted at all, and the cleaning out of the boxes, which is necessary in the zinc process, is only required at very long intervals. The lead, which contains from 2 to 12 per cent of gold, is then melted into bars and cupelled. The consumption of lead is 750 lbs. per month, equal to 3 cents per ton of tailings. The working expenses, including filling and discharging tanks, come to 80 cents per ton, which is divided as follows: Filling and discharging, 20 cents per ton; cyanide, 12½ cents; lime, 2.4 cents; iron, 4.4 cents; caustic soda, 10 cents; lead, 2.2 cents; natives' wages and food, 3.8 cents; coal, 9.2 cents; white labor, 9.2 cents; stores and general charges, 6.5 cents; total, 80 cents. At the Worcester works 100 tons are being treated per day; when working on a large scale it is anticipated that the expenses will be further reduced." (The Mining Journal of London, October 27, 1894.)

The advantages claimed for this process are that electrical precipitation is independent of the amount of cyanide or caustic soda contained in the solution, therefore in the treatment of tailings very dilute solutions can be used; for generating the current necessary in a 3,000-ton plant, 2,400 Watts are required, equal, theoretically, to 3½ horse-power, and actually requiring about 5 indicated horse-power. The process has produced 755 ozs. of gold during the month of July of this year.

(d) **The Pielsticker Process** applies likewise the electric current as the precipitating agent. A description of this process, illustrated by a diagram, will be found in the patent-specification (see Appendix). No information in reference to practical results could be obtained. The process attained of late a certain notoriety on account of the patent litigation now pending between the owners of the MacArthur-Forrest patent and the Cyanide Gold Recovery Syndicate (Limited) of London, who control the Pielsticker patent.

(e) **The Moldenhauer Process** of bullion precipitation consists in the application of aluminium, or alloys, or amalgam thereof, in the presence of a free alkali. It is claimed that aluminium separates the gold very quickly from the cyanogen solution without entering into combination with the cyanogen, but simply reacting with the caustic alkali which is present at the same time, forming therewith an aluminate. The precipitation of gold by aluminium takes place as follows:

$$6\ AuK\,(CN)_2 + 6\ KHO + 2\ Al + 3\ H_2O =$$
$$6\ Au + 6\ KCN + 6\ HCN + 6\ KHO + Al_2O_3$$

and

$$6\ Au + 6\ KCN + 6\ HCN + 6\ KOH + Al_2O_3 =$$
$$6\ Au + 12\ KCN + 6\ H_2O + Al_2O_3.$$

The whole of the cyanide of potassium which has been combined with the gold is being regenerated, and the consumption of the cyanide is limited to the loss involved by such secondary reactions as act decomposing. (See "chemistry of the process.") The discoverer of this method of bullion precipitation claims that the quantity of aluminium required for precipitating the same quantity of precious metal, is about four times less than the amount of zinc required to produce the same effect. No results of this process applied on a large scale have as yet been made public.

(f) **The Johnston Process** of abstracting gold and silver from their solutions of alkaline cyanide (United States patent 522,260, see Appendix) consists in the use of pulverized carbon, preferably in the form of charcoal. "The pulverized carbon is placed upon suitable supports so as to form it into filters, through a series of which the cyanide liquid is caused to pass successively, leaving the metal deposited upon the carbon. The gold and silver are then recovered by carefully burning the carbon and smelting the residue with the usual fluxes. By thus employing a series of filters, through which the solution is passed successively, 95 per cent of the precious metal contained in the solution is recovered. When only one filter is employed, only about one fourth of the gold can be extracted."

V. PERCENTAGE OF EXTRACTION.

The percentage of extraction depends on the character of the ore. As I mentioned before, the process is suitable for many ores which for chemical and mechanical reasons are refractory. The commercial question in the selection of a metallurgical process for treatment of a certain ore has to be considered parallel with the chemical, and that process should be adopted which permits the extraction of the largest percentage of bullion at the lowest cost, and with the least investment of capital. The cyanide process is, for this reason, the best yet discovered for the treatment of the tailings of the South African gold fields, although giving only an average of extraction of about 70 per cent, of which about 60 per cent is recovered (see page 52). No other process gave, at the same expenditure, any results approaching it. The conditions of the Witwatersrand ores are considered particularly favorable for the process, yet the extraction figures are, in most instances, not high. The percentage of extraction in various mills in Johannesburg will be given in the chapter on the process in Africa, page 60. The ores and tailings in New Zealand, where cyanide treatment of dry-crushed ores is carried on extensively, give better results. The Waihi ores, pure quartz, the gold free, but exceedingly fine, the silver in form of sulphides, no sulphurets of base metals, give an extraction of from 85 to 91 per cent of the gold assay-value, the silver returns varying from 43 to 51 per cent. The ore of the Crown mines, which resembles those of Waihi, but containing occasionally telluride of gold, yields on an average 93 per cent of gold and 79 per cent of silver. Concentrates, if satisfactory at all in cyanide treatment, give as a rule very high figures. A considerable quantity of concentrates from the Sylvia Mine in New Zealand, of a very complex character, being composed chiefly of zinc-blende and copper pyrites, with a large percentage of galena and iron pyrites, were treated by me by cyanide, and gave very satisfactory results under conditions where no other means of treatment were at disposal. The said concentrates are classified by the dressing plant; the fine slimes rich in bullion and galena gave as high an extraction as 95.43 per cent of the gold and 86.69 per cent of the silver; coarse concentrates gave an average of 80.32 per cent of the gold and 50 per cent of the silver. A large parcel of very fine sulphurets (from the canvas plant) from the Utica Mine, California, consisting of pure iron pyrites in finest division, mixed with more or less fine sand and carbonate of lime, proved an excellent material for cyanide treatment; the extraction averaged 93.18 per cent of the gold value, rising in some instances as high as 96.57 per cent. The coarse concentrates from the Frue vanners did not give such good results, if treated direct; their reduction to greater fineness, however, improved results. An appended table shows the results of successful treatment of parcels of ores from various sources. It is to be regretted that no corresponding table, giving a like description of ores treated with unsatisfactory results, can be produced for comparison, which would be useful and instructive.

The recovery of the bullion should correspond with the extraction shown by assays; in practice, however, there is often a discrepancy, which may be explained by various causes; new vats, particularly those of wood, absorb both gold and cyanide, and considerable differences in the returns will be felt during the first weeks of their use. In the Waihi Company's works in New Zealand, for instance, 116 tons of ore, of an assay-value of about $25 per ton, returned after the first month only 75 per cent of the gold, instead of 85 per cent as shown by assay; the returns of the second month yielded 80 per cent, instead of 91 per cent; after the third month the actual results came up to the extraction, as per assay—89 per cent. Similar experiences have been made in the Sylvia Company in New Zealand and elsewhere. It has been recommended to soak the wooden parts of a new cyanide plant with paraffine to prevent absorption; a coat of asphalt dissolved in bi-sulphide of carbon will be found a good preventive for the absorption by, wood. The chief sources of chronic losses are to be found in the imperfect separation of the gold solution from the exhausted ore residues, and in the faulty methods of dealing with the bullion after its precipitation by zinc. There is no reason why the actual returns should differ from the returns as established by assay, provided all mechanical losses are prevented. In reference to the losses in the Johannesburg mills, see chapter on the process in Africa, page 52.

VI. WORKING COSTS OF THE PROCESS.

As may be deduced from the whole tenor of this paper, the working costs of the cyanide process vary within wide limits and depend on many circumstances. Locality is a prime factor in the costs of working any process, and expenses must be high where operations have to be carried on in an inaccessible situation, or where there is dearth of fuel, water, building material, etc. Apart from the question of locality, the cost depends principally upon three factors:

The nature of the ore.

The price of labor.

The price of cyanide.

When an ore contains acid salts and demands an alkali treatment, the price of the alkali must necessarily be added to other costs; and where the ores are slimy, recourse must be had to drying and mixing appliances, which also increase the cost to an extent depending on local circumstances. The principal labor involved in the process is the charging and discharging of the vats, or, if agitation is used, the charging of the agitator and the removing of the exhausted material from the filter appliances. The charging and discharging of the percolation vats may, under ordinary circumstances, be contracted for at a rate of about 25 cents per ton; the extent of the operations is naturally of great influence in regulating the cost of handling. Very large works apply mechanical means for discharging the vats, such as dredges and movable cranes, which reduce the expense of labor per ton of ore to a minimum. To give an instance of the labor employed in working the agitation process I mention the Utica works in California, where the handling of the ore and all the labor connected with the treatment amounts to $1 per ton; this applies to concentrates varying from $50 to $95 in value per ton. The cost of the cyanide is one of the principal charges in the process, and the cost of treatment depends to a great extent upon the price of cyanide and on the amount consumed per ton of ore. The price of cyanide of from 95 to 98 per cent strength now averages about 50 cents per pound, delivered at seaports, and for lower strength the rates are somewhat better.

The amount of cyanide consumed per ton of ore is between 1 lb. and 3 lbs.; the character of the ore has, however, the greatest influence on the consumption, and in many cases the cyanide process will be found the best, cheapest, and quickest method, even if a considerably larger amount of cyanide is consumed per ton. Naturally, as the quantity treated is greater the cost becomes proportionally less. In some mines, as at the Primrose

Company, Johannesburg, tailings are treated for about $1 per ton, and it is very seldom that the cost in the Transvaal exceeds $2 50. In the Crown Reef works the cost of treatment ranges from $1 to $1 37 per ton; this includes the royalty, which for the use of the MacArthur-Forrest patents amounts in South Africa to $1 25 per standard ounce of gold. In Revenue, Montana, the cost of treatment per ton of ore, including crushing, amalgamating, cyanide treatment, and royalty of $1 per ton, amounts to from $4 to $5 per ton. The cost of treatment of ore at the Mercur Mine, Utah, amounts to $2 40 per ton, not including royalty. The cost of ore treatment in the Crown mines, New Zealand, is from $3 37 to $3 50; in the Waihi Company's works, New Zealand, the cost amounted to $2 25 per ton of ore; the expenses are now reduced to $1 25, cost of crushing and patent-royalty of 7½ per cent on the bullion value not included. The treatment of concentrates is, as a rule, more expensive than that of ore and tailings; their value is, however, in most cases, considerably higher than that of those materials, so that the cost per ounce of gold extracted is, with concentrates, generally much lower than with ore and tailings. The agitation treatment of concentrates (sulphurets) costs in the Utica works, California, from $3 25 to $3 50 per ton, labor included. "Ores yielding upwards of 90 per cent of their gold assay-value have been treated at $1 25 per ton, and tailings containing less than $3 have been worked profitably. It is therefore safe to assume that under favorable circumstances, and apart from all costs of mining and crushing, the cyanide process is capable of application at a low figure."

VII. COST OF CYANIDE PLANTS.

The cost of cyanide plants varies naturally with the system applied and the extent of the works. A well-equipped plant with a capacity of 50 tons per day will cost about $25,000, a 100-ton plant about $40,000. I shall have occasion to give details on the cost of plants when describing prominent and successful plants in different parts of the world, and I refer particularly to the corresponding chapter of the process in Africa, page 55.

VIII. MACHINERY AND APPLIANCES.

In discussing the various methods of applying cyanide, the machinery for each purpose has been described (see Chapter IV). Generally speaking, all plants have the same main features; modifications, however, will be suggested by special conditions of locality and character of ore. Economy in handling the ore is of the greatest importance, and should be made of first consideration in selecting the site for the plant and in its arrangement. The crushing machinery, if ores are to be treated without previous amalgamation, should be selected in accordance with the character of the material. The proper preparation of the ore is a very important item, and the crushing machinery should be selected so as to produce the smallest amount of dust if dry-crushing and of slimes if wet-crushing of the ore is practiced. For dry-crushing, rolls should be preferred, on account of their giving a product of greater uniformity than stamps, which are now used to a great extent. It is with the cyanide process as with other leaching processes, the more equal in size the particles are, the better. Wet-crushing, with the improvements necessary for mastering the slime difficulty, may ultimately win. Pumps of all constructions may be used for the transportation of the solutions, provided their material is not attacked by the alkaline cyanide solution. I shall refer to the general arrangement of plants in the corresponding chapter of the process in Africa, page 55.

IX. LABORATORY WORK.

Exact laboratory experiments must precede all cyanide mill operations; the required fineness of the ore, the strength of the cyanide solution, and the length of time for its action on the ore, have to be established by experiment. The correct strength necessary for treating any class of ore may be readily determined in the laboratory by treating a weighed quantity of the ore with cyanide solutions of different strength and for various periods of time. After treatment, the amount of gold extracted and the quantity of cyanide consumed should be determined. These results are then compared with the original contents of gold and silver in the ores and the original strength of cyanogen in the solution used. A method of rapidly determining the gold in the cyanide liquors consists in evaporating a known quantity to dryness on lead foil (free of silver), and cupelling the lead in the usual manner. In the presence of base metals, the liquor should be evaporated with the addition of litharge, and the residue assayed for gold and silver in the usual manner. The point to be aimed at is to consume as little as possible of cyanide and to extract at the same time as high a percentage of gold and silver as possible. The water used for making up the solutions should be examined for carbonic acid, free sulphuric acid, and sulphates. A chief point to be investigated in the laboratory is the " acidity " of the ore, by which term is understood the presence of products arising from the decomposition of sulphurets. These chiefly consist of free sulphuric acid and of products derived from a more or less advanced oxidation of pyritic matter, such as proto- and per-sulphates and basic iron salts. The exact amount of free acid contained in an ore sample can be readily determined by shaking a certain weight with water, and adding standard normal or one-tenth normal caustic soda solution, till the point of alkalinity is reached, as indicated by litmus or some other indicator. The means to prevent the ill effects of acidity have been discussed in the chapter on chemistry. The amount of soda or lime required for the ore is easily calculated from the consumption of normal soda solution as shown by the above experiment. *The cyanide solution will in many cases contain after the treatment of ore the evidence of secondary reactions, and a complete chemical analysis of the solution, before it comes into contact with the zinc, should in all cases be made. The explanation of unsatisfactory results with cyanide treatment will be found in many instances by means of such an examination.*
The reasons why one ore yields its gold readily to cyanide treatment, and others of a similar chemical composition do not, are not always apparent; chemical analysis and microscopical examination should be used along with practical tests. As a programme for the examination of an ore in reference to its fitness for cyanide treatment, the following may be suggested: (1) The ore shall be crushed and passed through a 30-mesh

sieve, and part of it assayed. (2) The "acidity," if any, of the ore, shall be determined, as described before, in say 100 grams; if necessary, water and alkali washes will then be applied before the ore is submitted to cyanide treatment. (3) Shaking tests in glass bottles with solutions of various strength during various periods of time will determine the amount of cyanide consumed by the ore, the strength of solution required, and the time necessary for the reaction. Cyanide determination in the solution and assay of the ore, after treatment, provide the required data. Agitation tests will decide if an ore is suited for cyanide or not. In cases of successful cyanide treatment by shaking tests, percolation tests of samples should then be made. The simplest apparatus for the purpose is a glass funnel of large size, the neck of which is closed by an india-rubber pipe and clip; a lamp chimney closed at one end by a stopper of india-rubber, which carries a glass tube, may be used for the same purpose. The ore is placed on a filter-bed of pebbles, which is covered by a piece of flannel or filter-paper. Experiments with equal weights of ore, but with solutions of various strength, and exposed during various periods of time, should be made simultaneously; analyses of the percolated solutions and assays of the well-washed residues will then show the best conditions of treatment. Small tests, if properly conducted, are excellent guides as to the treatment of ore on a large scale. In a cyanide plant there is ample employment for the chemist in charge; all operations should be controlled by him by assays and analyses; improvements on all parts of the process may be effected by careful observation and continued investigation. The non-success of the cyanide process in various mining camps may be attributed to the incompetency of the men to whose care the works were intrusted. Mine owners will find it profitable to employ, at least during the first time of their using cyanide, a competent man, instead of "finding out" for themselves at the expense of much time and money. When the process is once successfully and firmly established, an intelligent foreman will easily acquire the necessary knowledge to test and make up the solutions.

X. DANGER IN WORKING THE PROCESS.

The deadly poisonous character of the reagent was once considered to be a great obstacle in the way of its successful introduction. However, in the first place, the solutions used are so diluted that the hydrocyanic acid evolved from them is of no consequence if the works are properly ventilated. For safety, as much as for practical reasons, all materials should be tested for acid before treatment, and, if necessary, neutralized. In the second place, there is no necessity for those working the process to come into contact with cyanide, solid or in solution; even properly conducted cleaning up does not require contact. Some, but very few, have a susceptibility for cyanide, and with them the most diluted solutions, if brought into contact with the skin, produce eruptions, which, although not dangerous, are itching and annoying; such men should not be employed in cyanide works. In case of bad ventilation, complaints of headache, faintness, and dizziness may be heard. In instances where circumstances require that hands and arms should be brought in contact with the solution, I found a coating of oil or coal oil (kerosene) an effective protection for the skin. The process is now used so generally, and on such a gigantic scale, that many hundred men are constantly employed in the works; considering the dangerous character of the cyanide, the number of accidents is remarkably small, and it may justly be said that no more danger is incurred in the working of the cyanide process, under ordinary precautions, than there is in working in establishments of chemical and metallurgical industries, where corrosive liquids and acids are constantly in use. Extraction of gold and silver by cyanide will compare, as to fatal accidents, very favorably with most chemical and metallurgical industries. It is, however, always well to instruct the men how to act in cases of emergency: Put the patient into a hot bath and apply cold water to his head and back. In cases of internal poisoning, vomiting by physical means, or by emetics, is advised. Freshly precipitated carbonate of iron, obtained by mixing equal quantities of sodium carbonate and ferrous sulphate, is recommended for internal use; the two chemicals should be kept on hand ready for compounding this antidote. If the poisoning is the consequence of inhalation of hydrocyanic acid, it is advisable to make the patient inhale a small quantity of chlorine gas, ammonia, or ether; rubbing with camphor-alcohol is recommended. It has been reported of late that Dr. Johann Antal, a Hungarian toxicologist, has notified the Hungarian Society of Medicine that he had found a perfect antidote for prussic acid in nitrate of cobalt; he quoted forty cases of its use with perfect success.

XI. EXEMPLIFICATION OF THE PROCESS—THE PROCESS IN VARIOUS COUNTRIES.

After having given the outlines of the process, and the various methods of its use, I now propose to describe its practical application in different parts of the world.

A. The Process in Africa.

The cyanide process has found its earliest application on a large scale in the Witwatersrand gold fields of Johannesburg, South African Republic. The use of the process has since been extended to the different gold fields of the republic, the output of which has gone on steadily increasing, rather by application of improved machinery and the MacArthur-Forrest recovery process, than by the opening up of new mines. In the following notes, describing the methods of cyanide treatment in South Africa, I give chiefly the general information obtained from Mr. J. M. Buckland, the General Manager of the African Gold Recovery Company of Johannesburg, and embody in them, at the same time, the special information obtained from other gentlemen in charge of prominent companies and works.

The Ores.—"At the Witwatersrand, from which at present some forty-five thousand ounces, or about 90 per cent, of the gold obtained in South Africa by the cyanide process is obtained, the ore may be classed generally as conglomerate, locally called 'banket' (almond-rock). This consists practically of quartz pebbles embedded in a quartzose matrix, which last carries the gold and a varying proportion of iron pyrites. There is also a small quantity of alumina in most 'banket,' existing in diverse forms of córundum and (combined with silica) of clay. The latter is the cause of much of the slimes formed during the crushing process, the cheap and profitable treatment of which by cyanide still remains a difficult problem. The iron pyrites almost invariably contain traces of copper and nickel as sulphides. Within 50 to 100 ft. of the surface, the 'banket' is more or less weathered, and the iron exists in an oxidized form; but below this depth the ore becomes paler in color, less friable, and consequently more difficult to crush, and iron pyrites takes the place of the iron oxide. The gold also is more difficult of extraction by amalgamation or by cyanide; but whether in this latter case the cause is chemical or mechanical is not yet determined. On the latter hypothesis, the difficulty would consist in the fact that the particles of gold are so imbedded between the cleavage planes of the pyrites as to prevent to a great extent contact with cyanide solution; and, on the former, that the gold exists chemically combined with certain constituents of the ore, such as sulphur and arsenic—the cyanide having presumably not sufficient affinity to act as a solvent by breaking up the bond of union.

"At the Sheba Mine, De Kaap District, the ore is a quartzite, containing a small amount of iron pyrites and talc. When crushed, there are formed, owing to weathering, sulphates of iron and free sulphuric acid, which react on the talc (silicate of magnesia), forming sulphate of magnesia, and this salt may be observed as an efflorescence on many of the tailings-dumps. This compound, like all others in which the 'base' is weaker than potash and the 'acid' stronger than hydrocyanic acid, causes decomposition and loss of cyanide (see chapter on chemistry). At the Barrett Company (Kaapsche Hoop), the ore is a very soft, decomposed talcose slate, containing a large proportion of hydrated iron oxide. Owing to the very fine nature of the gold the solution of it is almost complete, but sufficient of the very finely divided portion of the ore passes through the filter cloth, coats the zinc, and prevents thereby the gold solution from coming into contact with it; precipitation is consequently seriously interfered with. The difficulty in such cases is met to a certain extent by the use of an extra vat, placed between the leaching vats and the precipitating boxes, in which the greater part of the suspended matter is allowed to settle, and tolerably clear liquid is filtered off. This ore is of so soft a nature as not to require any explosives in mining; it does not even require preliminary crushing for the purpose of preparing it for treatment, but is simply sifted through drums, and the portion which passes through the sieve, when mixed with coarse tailings to assist filtration, is ready for cyanide treatment.

Ore Reduction.—"At present, with the above-mentioned exception, all ores in the Transvaal gold fields, which are treated by cyanide, have previously undergone the ordinary crushing and amalgamation process in the battery. The wet-crushing stamp-mill is the only machine employed for ore reduction, apart from various grinding machines, such as the Wheeler and Berdan pans, which are rarely used for the finer reduction of tailings and concentrates. At the Rand (Witwatersrand), the average mesh employed for battery screens is thirty holes to the linear inch (No. 30 screen), or 900 to the square inch. In some cases, however, No. 24 screens are used. At mines, such as the Sheba, where the gold is very fine, No. 40 screens are employed, but the tonnage per stamp per twenty-four hours is in such cases not much more than two tons, as compared with the three and two thirds tons usual for mills crushing Rand 'banket.'

" In an experimental way, a solution of cyanide has been used instead of battery water; at present, however, water is invariably used in the mortar boxes, sufficient success not having attended the other method. The use of cyanide solution in the mortars would be of advantage only when the pulp is directly delivered into the percolation vats; the formation of slimes is fatal to this method. There is now no other process than the cyanide process used in treatment of tailings, and no mill of which the tailings are of sufficient value is considered complete without a cyanide plant. On the Witwatersrand alone there are upwards of forty cyanide plants in operation, and ten in process of construction; the present quantity of tailings treated monthly is about 250,000 tons. On the other gold fields—DeKaap, Lydenburg, and Klerksdorp—there are in all ten cyanide plants in operation.

Method of Applying Cyanide.—"The method of applying cyanide to gold extraction is that specified in the MacArthur-Forrest patents, of which the African Gold Recovery Company holds the rights in Africa, and lets out the right of using the process to gold-mining companies on royalty, usually 10 per cent on the value of the gold produced. Nearly all companies using the process erect and work their own plants, paying royalty as above." Some custom-works will be mentioned later on.

The mode of procedure in applying the cyanide process in Africa is generally that which has been outlined in the earlier part of this paper. Some further details, however, will here be given and may prove of interest. The general system of treatment is that of percolation in tanks.

The Vats.—"The percolation tanks, or leaching vats, vary greatly in size and shape. Those first constructed were from 15 ft. to 20 ft. square, and from 4 ft. to 5 ft. deep, the material used being boards 9 in. wide by 3 in. thick and as long as the side of the vat. Owing chiefly to the difficulty of making these water-tight, oval and finally round vats, composed of staves 2½ in. thick, 6 in. wide, and varying in length from 5 ft. 6 in. to 11 ft., were used. Vats now vary in diameter from 15 ft. to 40 ft., and in capacity from 30 to 600 tons. The employment of bottom discharge, by which the exhausted tailings, or 'residues,' are shoveled through a hole in the bottom directly into a truck below, has rendered the great increase in depth possible. 'Side discharge' through doors in the side of the vat is also employed to some extent. Discharge over the side into trucks on tram-lines is now used only in cases when first cost, or 'want of fall,' in the dumping-ground is a serious consideration. The loading into trucks does not cost more than 4 cents per ton, even under unfavorable conditions, and is only 2 cents and even less per ton in some cases. At the Barrett Company, the residues are shoveled through an opening in the bottom of the vat into a launder, where a stream of water carries them away." Some companies, like Le Champ d'Or French Gold Mining Company, let the filling and emptying of their cyanide tanks by contract.

"In some of the largest works cement vats are used, particularly for 'sumps,' or tanks, where the solution is stored after passing through the precipitation boxes. Such vats are really excavations lined with bricks, laid in hydraulic mortar, and plastered inside with cement; these attain a capacity of 600 tons, being 50 ft. in diameter and 9 to 11 ft. deep. For discharging them, when they are used as percolation vats, either tram-lines are laid down along the bottom, passing out through doors, which are bolted and made water-tight when the vat is in use, or else trucks are lowered by a steam crane into the vat, and filled by natives, and again hoisted" (see diagrams, page 57). "At the Salisbury works a tailings wheel lifts the pulp, after passing over the amalgamating plates, to a flume, which carries it to a hydraulic separator, which separates the slimes from the coarse tailings. This is done to render the subsequent treatment of the tailings more economical, as easy filtration is achieved when tailings free from slimes are treated. The slimes are treated by themselves and filtered in filter-presses. Many companies now run the material from the mill into intermediate settling-vats, provided with bottom or side discharge, for convenience of loading the trucks which transport it to the cyanide works. The difficulty of insuring an equal mixture of fine and coarse tailings is met by means of a rotary distributor, pivoted above the center of the vat and discharging into it. This distributor consists of pipes of different lengths diverging from a central basin into which the pulp is delivered. The openings at the end of the pipes are so arranged that the stream of pulp issuing therefrom causes the distributor to rotate." In the Nigel Company's works the slimes are separated from the sand by means of tailings-pits, with overflow into pits where the slimes are collected; these are dried, broken up, and delivered, mixed with the clean sand, into the vats in the works.

Treatment of Concentrates.—"Concentrates are not now treated by agitation in the Witwatersrand gold fields; percolation has been substituted for it. A period of contact and percolation, extending from two to four weeks, is now usually employed. Agitation has been abandoned on account of its cost, consequent upon the power and constant attention required, and the necessarily small amount treated at one time. In addition, it was found that the cyanide consumption was usually increased by the solution becoming heated, owing to the friction of the solid particles during an agitation of several hours. In treating by ordinary percolation, the concentrates are usually mixed with a sufficient amount of coarse tailings to insure filtration. Transferring the material from one vat to another at intervals of a few days is sometimes considered beneficial, for the purpose of obtaining a supply of oxygen as required by Elsner's equation." (See chapter on chemistry, p. 16.)

"Although pyrites themselves consume practically no cyanide, the great difficulty incurred in the treatment of concentrates generally, and of some tailings, is due to the fact of their having been partially oxidized by exposure to the air." The reactions thereby taking place have already been mentioned as detrimental in the chapter on chemistry, and the remedies enumerated (see p. 18). "The water-washings employed to remove the 'acidity,' as it is termed, take place in a special vat, as the traces of the cyanide retained by the filter-cloth, etc., of the regular leaching vat are liable to dissolve gold, and thus cause loss. Experiments have shown that a water-wash pure and simple will dissolve out of most tailings a minute quantity of gold, but this is so small an amount that it may be neglected. If the ore is not very acid a solution of caustic soda is run on after the last water-wash, and the air contained in the solution will serve to convert the ferrous hydrate formed, which would otherwise subsequently form potassic ferrocyanide

4cp

with the potassic cyanide, into innocuous ferric hydrate. If very acid, however, aëration, by changing to another vat, will be necessary.

"Lime sprinkled on the surface of the charge of ore, or mixed with it, is often preferred to caustic soda, and has the advantage of clarifying solutions from organic compounds, which, if present, cause 'frothing' in the zinc boxes. Lime does not form the yellowish-white precipitate in the zinc boxes, which is mainly ferrocyanide of zinc, and liable to occur when caustic soda is used, and which, by coating the zinc, interferes with the proper precipitation of the gold."

The Advisability of Ore Concentration.—"The question as to whether it is advisable to concentrate, or to allow the pyrites to remain with the tailings for subsequent cyanide treatment, is at present under discussion, and, like all other matters in gold extraction, resolves itself into a question of cost. The general rule is, that with high-grade pyrites concentration does pay, but with low-grade not." The Nigel Company has abolished the use of concentrating machinery. They found that the extraction of gold by cyanide treatment is equally as good from the ore from which the concentrates have not been taken, as it was when using Frue vanner concentrators. Their tailings are consequently of a high grade, containing pyritic matter, and solutions of greater strength are used than is the practice with other works on the Rand. (W. A. Radoe.)

The Cyanide Solutions.—"The strength of the solution before treatment was some four years ago ½ and 1 per cent, but now 0.25 and 0.3 per cent may be taken as the usual amount of pure cyanide of potassium contained in what are usually called strong solutions in South Africa. In most works a constant quantity of this strong solution is run on each charge of ore, having been made up from the 'weak' or dilute solution in stock by addition of a sufficient quantity of the solid salt, or a concentrated solution prepared from it. Should the cyanide consumption of the ore increase, the strength of the dilute solution, or that which has been already used, decreases, and more solid cyanide is required, and *vice versa*. Another method consists in always adding the same amount of solid cyanide to the same amount of weak solution, and in case the latter is below a certain point (say 0.1 per cent) to continually use a larger quantity of strong solution for running on the ore until the 'weak' rises to 'normal' once more. If the strength of the stock solution falls too low, the precipitation of gold is imperfect, probably because the cyanide of zinc formed in the precipitation boxes is not dissolved and coats the zinc. If, on the other hand, it is too high, the consumption of cyanide and zinc, by dissolution of the latter in the former, is unnecessarily great. In this case, too, the loss of cyanide by atmospheric decomposition is increased, and, while the same absolute amount of solution is lost by leakage and in the form of moisture adhering to portions of the residues, yet, the solution being stronger, more potassic cyanide is lost. The quantity of strong solution employed per charge of ore varies according to whether a preliminary washing with a dilute solution has been employed or not. In the former case, it is about 25 per cent of the weight of the ore, and in the latter case about 40 per cent, which last quantity is usually sufficient to just cover the charge. The amount, however, varies in different works, and, within reasonable limits, it is not a matter of great importance, provided sufficient solid cyanide is added daily to keep the stock of weak solution at the right strength. It is desirable that the strong solution be of uniform strength throughout the whole charge of ore, and this object is attained, in great measure, by using a preliminary washing with a dilute solution; the cyanide of the latter satisfies most of the components of the ore, which consume cyanogen, so that the strong solution, which follows, is free to act on the gold alone. This preliminary wash has also the advantage of saturating lumps of slimes which may be in the tailings and would absorb the strong solution, which would be lost when the residues are discharged. Generally speaking, a larger quantity of the weaker solution is preferable to a smaller quantity of strong solution, but exigencies of time, capacity of the plant, filtering properties of the material, etc., cause modifications of this rule. When a charge has had more than its own weight of washings passed through it, it becomes a question whether there is sufficient increase in yield by continuation of the process to cover the cost of pumping, apart from the fact that if the solution be run too fast through the zinc boxes not only is the gold it contains imperfectly precipitated, but that already deposited is liable to be mechanically carried into the sump by the force of the current."

The strength of cyanide solutions used in the Crown Reef Mine works varies from 0.05 per cent to 0.35 per cent; they range after treatment from nothing to 0.33 per cent. The total quantity of solution used, inclusive of water-washes, is about 80 per cent of weight of charge; extraction takes from forty to fifty hours. (G. E. Webber, Jr.) In the Nigel Company, where, as already stated, concentration has been abandoned, the conditions are in consequence somewhat different from those usually prevailing in cyanide works; for tailings over 24 dwts. (about $18) in value a solution of 0.6 per cent is used, preceded by a weak wash of solution of 0.15 per cent and followed by two weak washes; the liquors drain off at a strength of from 0.4 to 0.25 per cent of cyanide, the first solutions draining off at a lower percentage than the last. The amount of cyanide used per ton is about 3.8 or 3.5 lbs. per oz. of gold recovered (3.5 lbs. of 76 per cent cyanide). (W. A. Radoe.)

Precipitation by Zinc.—The precipitation process going on in the zinc boxes has been fully discussed in a former part of this paper.

" After passing through the zinc box the solution should not contain more than 50 cents of gold per ton, and in the majority of cases there will be only a trace of gold present. If appreciable quantities of gold remain unprecipitated, a certain amount is daily lost in the dilute cyanide solution contained in the residues (see above). The latter should be periodically tested for gold soluble in water, and gold soluble in cyanide solution; imperfect precipitation will be discovered by the first, and too short time of treatment by the second test."

Time of Treatment.—"The total time employed in the treatment of a charge of tailings varies from three days to a week, and is dependent, from a chemical point of view, upon the greater or less fineness of the gold; the general rule is that the longer the time the better until the increased cost of treatment more than counterbalances the improved percentage of extraction."

Cyanide.—The cyanide usually employed contains from 70 to 80 per cent of pure potassic cyanide, but another quality, containing upward of 95 per cent, imported from Germany, is also used, and is preferable for the reasons explained in the chapter on cyanide (p. 30). The consumption of cyanide is about 150 tons per month in the Witwatersrand mines. Germany has sent out nearly 1,000 tons to the Transvaal this year.

Value of Rand Tailings and Percentage of Extraction.—"The average value of Rand tailings per ton, before treatment, is $5; of this $3, or 60 per cent, is actually obtained by the cyanide process, $1 50, or 30 per cent, is left in the residue, and 50 cents, or 10 per cent, is unaccounted for. Ores from DeKaap, containing a large amount of mispickel, gave only an extraction of 9 per cent, but on roasting the extraction rose to 83 per cent, the arsenic being presumably driven off. Owing, however, to the cost of fuel, and the high cyanide consumption resulting from sulphates formed by partial oxidation, roasting is never employed as preliminary to the cyanide process. The percentage of gold extracted varies in different localities, but is usually between 70 and 80 per cent. It depends chiefly upon the degree of fineness of the ore, and the degree to which the gold it contains has been liberated from the matrix and exposed to the action of the solution. (In comparing an ounce of like particles of $\frac{1}{30}$ in. diameter with an ounce of particles of similar shape but $\frac{1}{40}$ in. diameter, the surface exposed by the first lot is three fourths of the surface exposed by the second lot—this being a particular instance of the general law, that for equal weights of similar particles the surface exposed varies inversely as the diameter.) There is little doubt that the remaining 20 to 30 per cent in the residues consist of particles of gold still incased in the matrix, and this is proved by the fact that finer grinding renders almost complete extraction possible. The limits of fine grinding on a working scale are fixed by the increased difficulty of filtration. Even when only a No. 30 screen is used in a wet-crushing stamp-mill, it is not possible to filter the pulp in its entirety, on account of the slimes; the consequence of this is that with moderately fine crushing only the coarser portion (possibly 80 per cent) of the tailings is at present treated by cyanide. Slimes of sufficient value are in some cases treated by drying, crushing, and mixing with a sufficient amount of coarse tailings to allow filtration. A mixture of equal parts of each take at least a week for treatment. The drying is performed either by exposure to the sun, or, especially in the case where much organic matter is present, by slightly calcining in a reverberatory furnace, or in form of bricks in a kiln. In the last two cases, however, the cyanide consumption is increased by the oxy-salts of iron formed, and although usually remarkably good results may be reckoned upon, yet the cost involved in so much handling is so high as to be prohibitive for low-grade slimes." The treatment of slimes or of tailings-pulp containing a high percentage of slimes, is still one of the unsolved problems of the cyanide process; at present most of the slimes are washed away, and with them a large amount of gold is lost in Johannesburg.

It has been suggested to solve the difficulty of slime treatment by mixing the slimes with 50 per cent of their weight of a solution containing cyanide of potassium and the double cyanide of manganese and potassium. This mixture is pumped into a filter press under high pressure; after filling the press, water is forced through, washing out the gold solution. An extraction of 97.6 to 98.2 per cent is claimed. (W. Bettel's process. E. & M. J.)

Loss of Gold, and its Causes.—"From leakage, and the loss consequent upon the handling of gold slimes in the various stages of conversion into bars of bullion, there is a certain amount of loss; but this, in properly conducted works, should be small. Even when experimental errors of assay be also taken into account, the discrepancy between theoretical extraction, estimated on assay and tonnage, and actual extraction of fine gold contained in the bullion, should not exceed 3 to 4 per cent. As a matter of fact, chiefly from unskillful work, the bullion actually recovered amounts in the Rand gold fields, on an average, to only 60 per cent; whereas the average extraction, as estimated by assaying the material before and after treatment, amounts to 70 per cent. The chief causes of error in the estimate of extraction arise from incorrectly calculating the tonnage, from faulty sampling of charges and residues, and from careless assaying of too small samples. If truckloads of tailings be weighed periodically and the percentage of moisture carefully tested; if samples of tailings before treatment be taken from each incoming truck and mixed thoroughly, and the residues after treatment be treated likewise; and if assays be daily made in duplicate on not less than an assay-ton of material, when this is low grade—then there should be little difference between estimated and actual extraction, provided there is not much leakage and the gold slimes are carefully handled. In many works a serious loss is incurred by allowing solution, containing very fine gold slimes in suspension, to enter the general stock of solution, and to be ultimately discharged with the residues. So fine is some of this material that it will even pass through the finest filter cloth and remain suspended after hours of 'settling.' The most effectual way of overcoming this difficulty is to run all solution, filtered or decanted, from gold slimes during the process of 'cleaning-up' or separating them from the zinc, into a separate vat, called a 'settler.' This is left undisturbed for some days, after which the supernatant liquid may be safely run off. This settler should equal in capacity the united zinc precipitation boxes, and may be cleaned up half-yearly. Filter presses are sometimes used to remove as much solution as possible from the gold slimes before drying the same." In large works, like those of the Crown Reef Company, the actual returns of bullion amount to 95 per cent of the calculated extraction. It is explained that the difference in the estimation of the weight of tailings treated is sufficient to account for the difference of 5 per cent (G. E. Webber, Jr.). The Nigel Company recover about 93 per cent of the calculated extraction; the incomplete recovery is attributed partly to the "soakage" of the vats, and particularly to the treatment of the bullion, the slag sometimes containing as much as two hundred ounces to the ton, of which only from 60 to 70 per cent are recovered by amalgamation in a grinding-pan (W. A. Radoe).

The Zinc for Bullion Precipitation " is used in the form of thread-like turnings, obtained as before described. A cubic foot of them weighs from 3 to 6 lbs., and exposes forty square feet of surface per pound weight. Granulated zinc is never used, as it exposes a very small surface in proportion to its weight and is liable to clog in the extractors. Aluminium in conjunction with an electric current has been suggested, as also alternate sheets of iron and lead foil between which a current of two hundred ampères and seven volts passes; but these methods are still in the experimental stage. The consumption of zinc varies greatly in different works, and is dependent upon causes other than the amount of gold precipitated. The absolute amount varies from 2 to 8 oz. per ton of ore, but of this, owing to waste in cutting out and turning, probably not more than one half goes actually into solution, when the finely divided zinc included in the gold slimes be also taken into account. The precipitation of the bullion is conducted as described before. The zinc consumption, above the amount required for gold precipitation by the equation of the chemical reaction, is due to its solution in the caustic alkali formed as indicated, in the free cyanide and caustic alkali present in the solution as it issues from the leaching vats, and also in its precipitating action upon other substances in solution. An average ore would probably consume about ten times as much zinc as it yielded bullion, but if in treatment caustic soda has been used in excess, the consumption will necessarily be higher."

Treatment of Precipitates—"Acid is occasionally used for making a complete clean-up of all the zinc contained in the boxes, and also for refining the amalgam (zinc, gold, and mercury) formed therein, when the tailings have contained much 'floured' quicksilver. Its use, for refining generally, is not advocated in Johannesburg, as it involves washing and filtration of the slimes, and loss of gold by the formation of regulus in melting, if sulphates have remained in the slimes by fault of imperfect washing. The method most in use for refining gold slimes in the South African gold fields is by the use of nitre. The slimes are dried till just before they become dusty; they are then mixed with powdered nitre, the amount varying from 3 to 33 per cent of their weight, and gently heated as a thin layer, either in a wrought-iron pipe (10 in. diameter by 6 ft. length), or preferably in a tray of wrought-iron ($\frac{3}{8}$ in. thick, by 6 ft. by 3 ft. by 1 ft.), which may also be used for the drying process. In neither case do the flames come into direct contact with the slimes; a hood carries off the obnoxious fumes. By the use of nitre everything in the zinc precipitation boxes which passes a sieve of three of four meshes to the lineal inch may be refined, and thus the finely divided zinc, which otherwise accumulates and clogs in the boxes, is constantly removed. Less nitre is always used than is required to oxidize all the base metals present, as otherwise the free nitre will rapidly corrode the plumbago crucibles, which subsequently are used for melting; it is advisable, however, to remain as near as possible below the limit, as the roasting which follows is thereby conducted quicker and at a lower heat. Besides rendering the bullion finer—containing say only 15 per cent base metals—this nitre-roasting gives a cleaner slag and lessens by at least one half the time required for fusing the gold slimes, and prevents violent ebullitions of vapor from the crucible. From 3,000 to 4,000 oz. of bullion can be obtained in twenty-four hours from roasted slimes containing 33 per cent of gold by the use of No. 70 plumbago crucibles, with good coke, in four box furnaces (20 in. square by 22 in. deep). The following fluxes have been found to answer well: When much metallic oxide is present—slimes six parts, borax four parts, soda two parts, sand one part. When little metallic oxide is present—slimes three parts, borax one part, soda two parts, sand one part. The function of the sand is to form a fusible slag with the soda, and also to protect the pots against metallic oxides and the potash formed by the reduction of the nitre. The slag resulting from melting slimes usually contains an appreciable quantity of gold. This, in the absence of smelting works, is generally crushed by hand in a mortar or by power in a smallest size Gates or Fraser & Chalmer's sample grinder. It is then panned, and the tailings resulting, still rich as a rule, are shipped to Swansea. In estimating the cost of a flux, it should be remembered that a very small percentage of gold in the slag will pay for an improved flux, and that flux which gives the cleanest, most fluid slag is preferable."

I have given here the mode of bullion treatment in Johannesburg as described by the African Gold Recovery Company; in addition to it, I refer to my own way of procedure, as given on page 36, which may offer some points of advantage.

Fineness of Bullion.—The bullion in the Robinson Works is about 650 fine; it is very hard and brittle, and the bars are by no means uniform, so that it is difficult to obtain an accurate assay; in addition to zinc, they contain silver, lead, and sometimes a little copper (Butters and Clennell). The bullion of the Crown Reef Company is 950 fine (830 gold, 120 silver) (G. E. Webber, Jr.). The bullion of the Nigel Company has been on an average 795 fine during the last seven months (W. A. Radoe).

The Cost of Treatment "varies according to the size of plant and the facilities for working, but exclusive of royalty may be taken at from $1 50 per ton for a 5,000-ton plant, to 87½ cents per ton for plants treating 20,000 tons monthly."

The Cost of Plants "varies according to the locality and the style in which they are erected. To erect an average size plant costs at Johannesburg about $6 25 per ton of ore it is intended to treat monthly; for very large plants the cost would perhaps be $5 per ton." All plants have the following main features: Leaching or percolation vats, zinc boxes for bullion precipitation, sumps or tanks for storing solutions, pumps for assisting filtration, and pumps for transporting the liquids. The difference between the various plants consists in the size, form, and material of the vats, the system of charging and discharging the tailings, and the general arrangement of the different parts of the machinery. The construction of vats and the handling of tailings has been discussed above. In reference to the latter, I attach details of discharging appliances, taken from "Notes on Gold Extraction," by W. R. Feldtmann.

The General Arrangement may be of different kinds. "The most convenient method is to have solution vats, leaching vats, extractors, and sumps in four tiers, so that each series may be completely drained into that next below it. By this means sufficient solution can be stored in the solution vats, and sufficient room left in the sumps to enable work to proceed for from twelve hours to twenty-four hours without pumping. Many plants, however, have the solution vats and sumps on the same level as the leaching vats; in this case the solution issuing from the last mentioned vats is run through precipitation boxes into a small tank and is continually pumped back when required." I attach plates, which illustrate the variations in the general design of plants with regard to the relative position of the different parts, which with their explanation are taken from "Notes on Gold Extraction by means of Cyanide of Potassium, as carried out on the Witwatersrand Gold Fields," by W. R. Feldtmann.

"In No. 1 design, the leaching vats are placed highest. The solution gravitates from these through the zinc boxes into the storage vats, there to be made up to strength ready for pumping up to the leaching vats again. In the sketch the discharging of the tanks is assumed to be done over the side. In No. 2 design the solution is either pumped direct from the leaching vat, or, running into a small sump or an air-tight receiver, is pumped from there into zinc boxes, and runs thence into

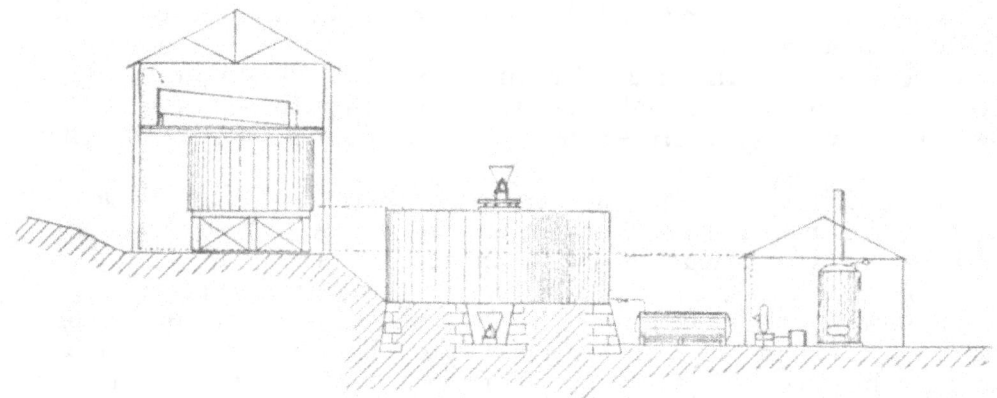

— VARIATION Nº 2 —
— IN —
— DESIGNS OF CYANIDE PLANTS. —

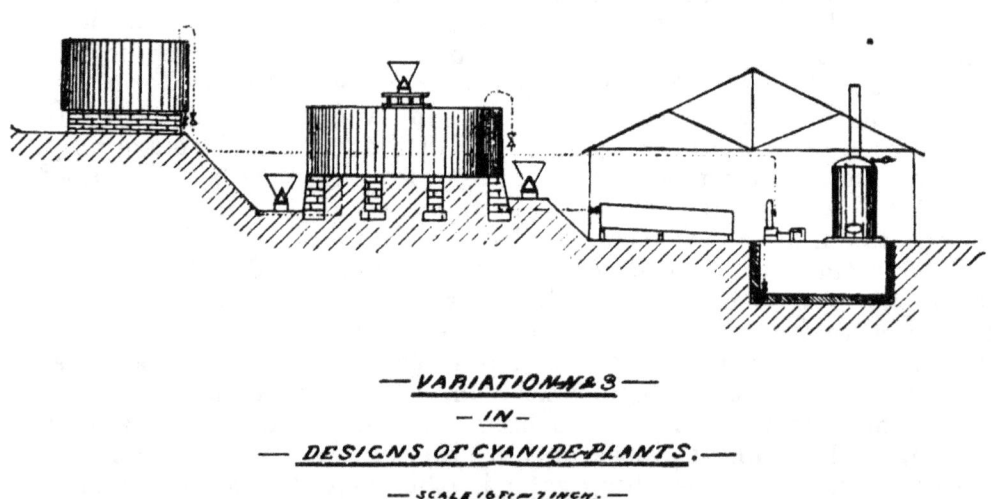

— VARIATION Nº 3 —
— IN —
— DESIGNS OF CYANIDE PLANTS. —
— SCALE 18FT = 1 INCH. —
COP. FROM W. A. FELDTMANN'S NOTES
ON
GOLD EXTRACTION "ETC.

20 15 10 5 0   10   20   30   40

overhead storage vats. Having been made up to strength, it is ready
to run direct into the leaching vats again. The discharge system indicated is the 'bottom discharge.' No. 3 design is a combination of the
two previous ones, and is advantageously fitted with a pipe service to
enable one, if desired, to run solutions up through the sand in the
leaching vats. As shown in the sketch, the plant is designed for side
discharge; but of course any system of discharge may be applied to
any of the three arrangements of plant.

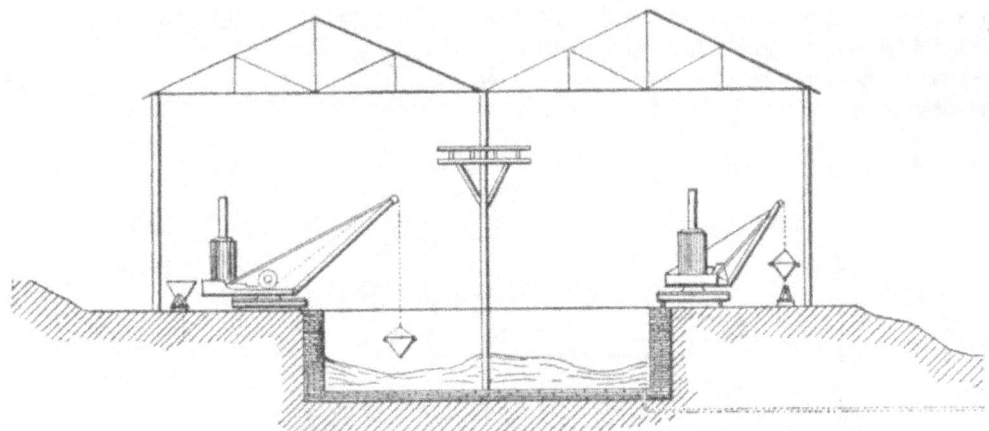

— DISCHARGING TAILINGS VATS —
— AT THE —
— LANGLAAGTE ESTATE Cos PLANT —
— SCALE 20 FEET = 1 INCH —

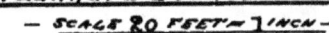

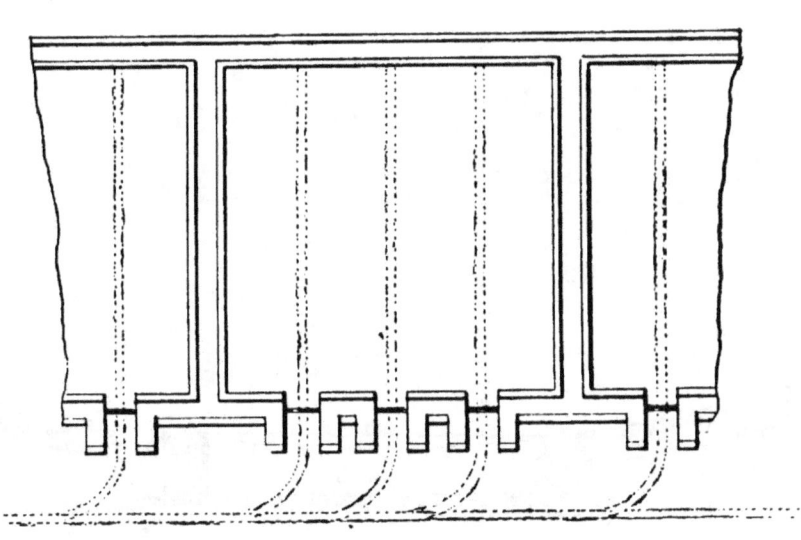

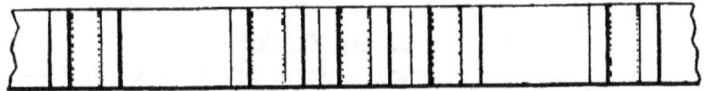

— SQUARE FILTER VATS AT THE WORKS OF —
— THE CROWN Cº —
— WITH DOORS FOR THE DISCHARGING TRUCKS. —
— SCALE 20 FEET = 1 INCH. —
— COP. FROM W.R. FELDTMANN'S NOTES —
" ON
GOLD EXTRACTION" ETC.

Buildings.—" The majority of plants now erected have only the zinc boxes inclosed in buildings, and there is little objection to having no weather protection when cemented vats only are used, but with wooden vats, exposure to the sun and weather undoubtedly causes increased leakage.

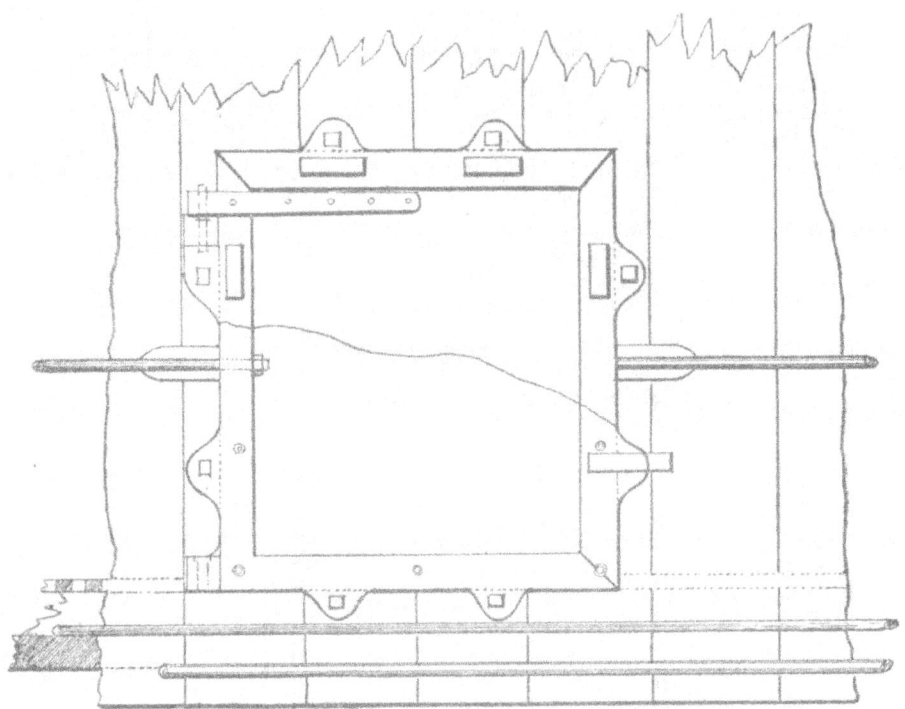

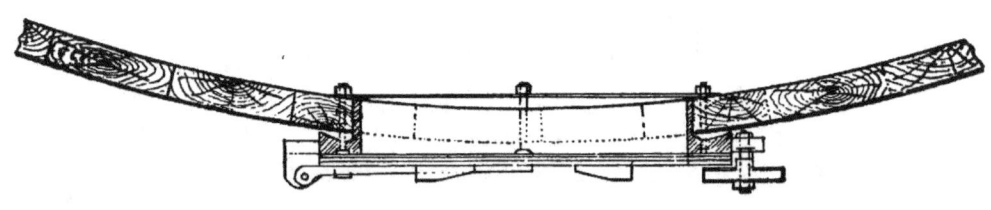

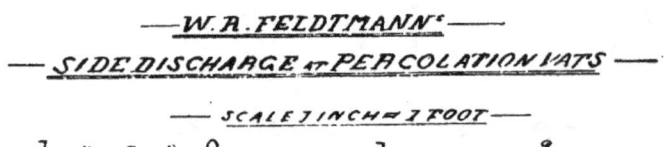

—*W. R. FELDTMANN:*—

—*SIDE DISCHARGE of PERCOLATION VATS*—

—*SCALE 1 INCH = 1 FOOT*—

Labor.—" The men employed in a plant of average capacity (5,000 tons per month) are: One manager, one assayer, two shift men, one mechanic, two native gangers, and the native crew." The Nigel Company employ two white men of twelve-hour shifts in the works, whose duty it is to "make-up," pump in, and drain off solutions, and to attend

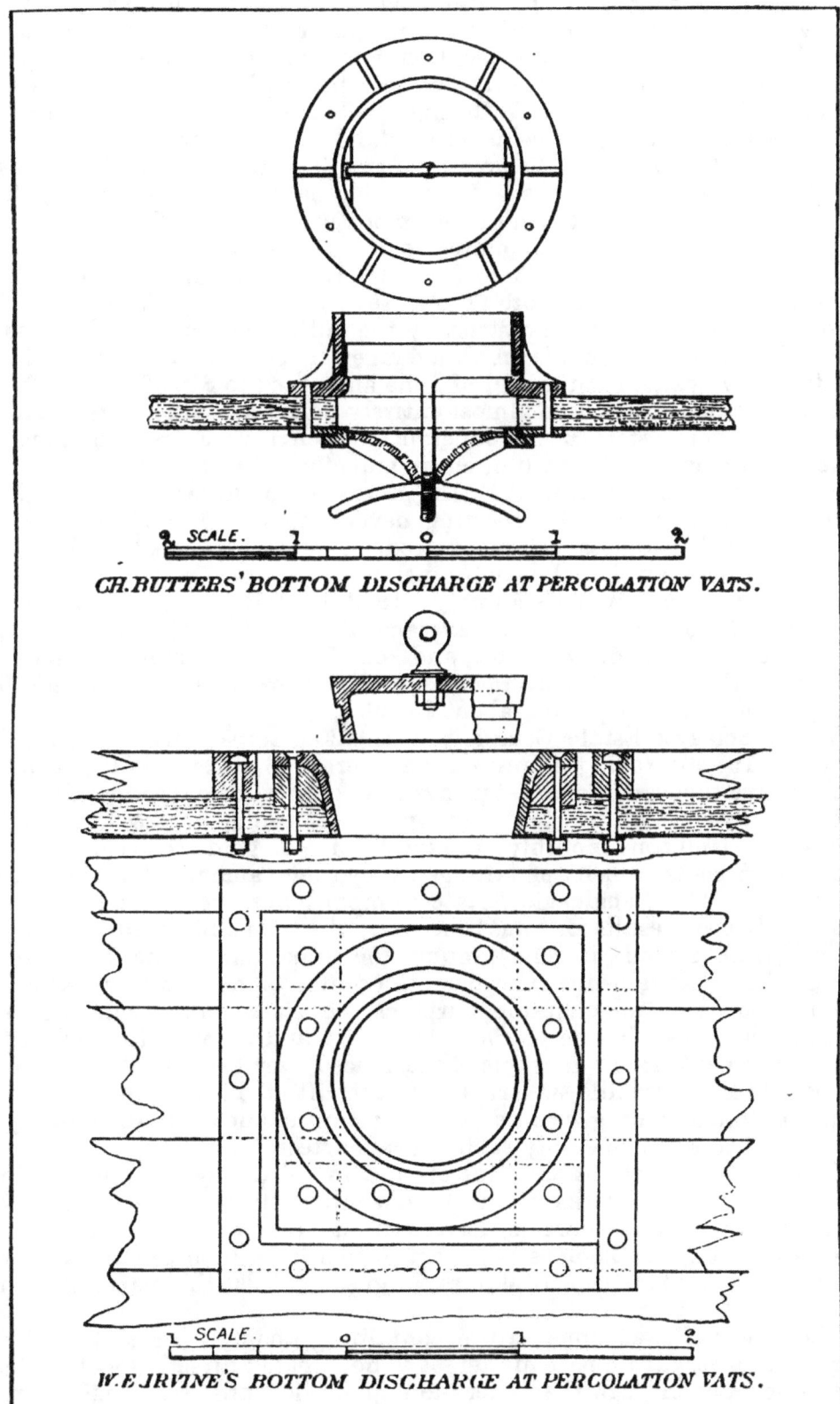

CH. BUTTERS' BOTTOM DISCHARGE AT PERCOLATION VATS.

W. E. IRVINE'S BOTTOM DISCHARGE AT PERCOLATION VATS.

COP. FROM W.R.FELTMANN'S "Notes on Gold Extraction" Etc.

to things generally, and one white overseer over the natives (about thirty in number), who attends to charging and discharging the vats (W. A. Radoe). The Crown Reef Company employ one white man per shift of eight hours and seventy native laborers, handling 500 tons of tailings (G. E. Webber, Jr.). During 1893 about 500 vats were in use on the South African gold fields, of a total daily capacity of 11,000 tons.

Results of the Process.—The process of treating tailings by cyanide was first introduced on the Transvaal gold fields in 1890; the output of gold by its means has since taken on enormous proportions. The gold produced by cyanide in 1890 amounted to 286 oz.; in 1891 to 34,862 oz; in 1892 to 178,688 oz.; in 1893 to 330,510 oz.; and during the first six months of 1894 to 317,950 oz. I inclose a table giving the output of gold in the Witwatersrand district by the milling and cyanide process. The value of cyanide gold is on an average $15 per oz. The total value produced by cyanide at the end of June amounted to $12,934,440. This enormous amount has been almost entirely derived from low-grade tailings, which were, before the introduction of MacArthur-Forrest's methods, practically valueless, for want of cheap and efficient means of extraction. The working cost of treatment by the process is the lowest known in the history of gold metallurgy, roasting never, and preliminary treatment in exceptional instances, being required. The process is adopted, only amongst others, by the following mines:

Robinson Company, 60 stamps; in 1892, 101,061 tons were crushed, yielding 98,799 oz. of gold, an average of 19 dwts. 13¼ grs. per ton; the concentrates yielded 8,407 oz., and 75,375 tons of tailings by cyanide treatment 27,577 oz.; the average from all sources was 1 oz. 6 dwts. 16 grs. per ton; the cost per ton about $5 50.

The Langlaagte Estate Company, with 120 stamps working, crushed in 1892, 197,201 tons, yielding on an average 6 dwts. 23.81 grs. per ton; twelve thousand tons of tailings are being worked per month at a working cost of $1 per ton; this plant is being increased to treat thirty thousand tons monthly; the total value of gold received, including that from treatment of tailings by cyanide, amounted in 1892 to $1,554,850. The Langlaagte Estate Company has been using cement tanks with good results for some time past; the vats and sumps are built in excavations made in solid ground, the upper part being level with the ground. A tram-line runs over the center of five large vats, each holding 450 tons, for charging purposes. A steam crane, running on rails, placed on each side of these vats, lifts and empties the tailings, after treatment, into a movable chute placed over two parallel lines of trucks. This chute fills two trucks at each lift at a cost of less than 2 cents per ton, including wear and tear and maintenance of the machinery. The time necessary to empty each vat is fourteen hours.

The following description of the application of the cyanide process in the very extensive works of the Langlaagte Estate and Gold Mining Company in Johannesburg is derived from notes written by the company's assayer, Mr. Thomas Lockhart, which have been contributed by the company's manager, Mr. James Ferguson, whilst this paper was in press:

The ore is of a silicious nature, containing about 2 per cent of iron pyrites, and very little base matter; it is not treated directly by cyanide, but is first put through the battery-amalgamation process. Eight hundred tons are crushed daily. The tailings leaving the plates are concentrated, and then run to three settling dams, each of which holds 7,000 tons. The slimes are here separated from the tailings and allowed to run away, as they are greatly impeding the percolation of the cyanide solution in the vats. The tailings, free from slimes, are hauled from these dams in trucks, by means of two endless steel wire ropes, and run onto an overhead tram-line, from where they are dumped directly into the vats for treatment. The vats are built in excavations; their tops are level with the surface of the ground. Of these vats there are fifteen; they are built of brick, and circular in form, of a diameter of 40 ft., and a depth of 9 ft. 3 in. Their sides are cemented, and the bottoms laid with concrete; the latter slope gently to one side, where a 2 in. pipe carries off the solutions. The bottoms are covered with a filter-bed, built of stones, about 3 in. deep, which are packed close together and form a level surface, which is covered with cocoa matting. Each vat holds about 450 tons of tailings, which remain in them under treatment for three days. After the completion of the extraction, the residues are discharged by means of trucks, which are lowered into the vats, filled by Kaffirs, hoisted up by a crane, and run out on the dump by means of an endless rope.

As a rule three vats are charged per day and three discharged. During March 33,000 tons of tailings were treated on 31 working days, or 1,064 tons per twenty-four hours. These tailings averaged before treatment 4 dwts. 5 grs. of gold per ton; the residues, after treatment, contained 17 grs., which corresponds with an extraction of 83 per cent, but only 68 per cent of the assay gold-value was actually recovered. The chief causes of the incomplete recovery of the gold are: The imperfect precipitation by zinc; the presence of slimes in the tailings, which take up and retain some of the cyanide gold solution; the loss of gold solution through leakages in tanks and pipes; and the mechanical losses in manipulating the precipitates in drying, calcining, and melting.

The strength of the first or strong cyanide solution pumped onto the tailings, varies from 0.32 to 0.38 per cent; it stands 3 or 4 in. above the surface of the charge, and remains in contact for from eight to twelve hours; after that time it is run off and passes through the zinc precipitation boxes; it is then brought up to its original strength by adding cyanide of potassium and used again on a next charge. The zinc boxes are 18 ft. long by 5 ft. wide and $2\frac{1}{2}$ ft. deep, and contain the usual divisions. Five such boxes take the solution from ten vats.

The first part of the solution, when running off the charge, contains about 0.02 per cent cyanide; the following portions are of greater strength and rise up to 0.3 per cent. When the solution has drained off, the second, or weak, solution, of about 0.2 per cent cyanide, is pumped on; this percolates directly through and replaces the first solution, which has remained in the material. A water-wash completes the treatment. This wash remains in contact during some hours, and is then drained off as usual; it forms the weak or second solution for a following charge.

The solutions are stored in vats 56 ft. in diameter and 9 ft. 3 in. in depth. The consumption of cyanide amounted during the last six months of 1893 to 0.55 lb. per ton of tailings.

The treatment of the concentrates (sulphurets) is conducted on the same plan as that of the tailings. Two tanks, each 35 ft. in diameter and $2\frac{1}{2}$ ft. deep, serve the purpose. A solution of 0.6 per cent cyanide is used; this remains in contact with the material for six hours; it is then run off and the gold precipitated as usual in the zinc boxes, which are 11 ft. long, 2 ft. deep, and 2 ft. wide. The solution is then restored to the required strength and used over and over again till the assays of the residues prove the termination of the extraction.

During the month of March 405 tons of concentrates were treated; they averaged before treatment 2 ozs. 10 dwts. 4 grs. of gold per ton; the residues contained 5 dwts. 14 grs., which corresponds with an extraction of 89 per cent of the assay-value; this percentage is actually recovered. The consumption of cyanide per ton of concentrates amounted during the last half of the year 1893 to 0.5 lb. The cyanide used contains 98 per cent of potassium cyanide. The zinc used for precipitating the gold contains 1½ per cent impurities; it is applied in filiform. For every one ounce of fine gold obtained 1.48 lbs. of zinc are consumed. Cleaning up takes place twice a month. The gold precipitates, mixed with finely divided zinc, are separated from the coarse zinc by shaking; they are then dried and calcined. The calcination is carried on on a heavy iron plate, which is heated to redness; it is continued till the oxidation of the zinc is complete. The calcined mass is then fused with borax, soda-ash, and sand. The mixture is charged into No. 50 plumbago crucibles and melted in a reverberatory hearth furnace, which holds twenty-two crucibles at a time. The time required for melting varies from one and a half to three hours, according to the character of the material and the temperature of the furnace. The molten mass is poured into iron moulds; the bullion, thus obtained, is remelted into bars of 600 oz. The fineness of the bullion varies considerably; that obtained from tailings runs from 700 to 780; that from concentrates from 750 to 820 fine. The weight of the fluxed precipitates which are reduced per month amounts to about 9,000 lbs. Three shifts, of two white men each, attend to the working of the process. Two twelve-hour shifts, of five Kaffirs each, are constantly employed in cutting zinc shavings. The charging and discharging of the percolation vats is done by contract; 14 white men and 250 Kaffirs are employed for the purpose. The buildings consist of wooden frame structures, covered on all sides with corrugated iron.

The Langlaagte Company pays a royalty for the use of the Mac-Arthur-Forrest patents, amounting to $1 25 per standard ounce of gold extracted from tailings, and 62½ cents per standard ounce of gold extracted from sulphurets.

The Ferreira Company derived for the half year ending March 31, 1892, a profit of $22,030 on the treatment of 15,310 tons of tailings, producing 3,495 oz. of gold at a cost of $2 91 per ton. For the year ending March 31, 1893, the tailings treated by cyanide yielded 11,201 oz. of gold, which means to say 4,592 dwts. of the value of $4 16 per ton.

The Crown Reef Company has 210 stamps working. For the half year ending March 31, 1893, the tailings treated by the process yielded 16,629 oz. of gold. The revenue from tailings and concentrates amounted to 92 cents per ton.

The Henry Nourse Company treated, during the first six months of 1893, 12,640 tons of tailings by cyanide, yielding 3,040 oz., or 4.81 dwts. of gold, at an average cost of $2 33 per ton.

The Meyer and Charlton Company treated during six months ending June 30, 1893, 10,799 tons of tailings by the cyanide process, yielding 3,205 oz., or 5.93 dwts. per ton.

The Nigel Company obtained during twelve months 21,471 oz. of gold by the process from tailings and concentrates.

The Rand Central Ore Reduction Company has been formed to purchase tailings from companies who have not erected their own plants; 1,500 tons of tailings are treated in their works per twenty-four hours. The vats, made of wood, hold from 75 to 600 tons. The tailings average 5 dwts. before and 1 dwt. after treatment. The strong solutions contain about 0.3 per cent, the weak about 0.15 per cent of cyanide. The last solution runs off at about 0.08 per cent. All solutions together amount to about 1 ton per ton of ore. The extraction takes about three days. The cyanide used is of 98 per cent on the average. The extraction from the tailings varies from 75 to 80 per cent of the assay-value, of which 80 per cent are actually recovered; the clean-up is never complete and much gold remains in the slag. The zinc consumption amounts to about 1 lb. per ounce of gold recovered. Electricity is also used for bullion precipitation. The bullion after melting is 750 to 850 fine. The total cost of treatment is about $1 per ton. The cost of the plant amounted to $600,-000; 200 white men and 800 Kaffirs are employed. (D. Ruston.)

The Langlaagte Royal Gold Mining Company is erecting a cyanide plant for the treatment of 10,000 tons per month. The mentioned companies, in conjunction with a number of others, more or less important, treat upwards of 200,000 tons of tailings monthly. During June, 1894, 220,507 tons were treated, with an average yield of $3 per ton. The treatment of concentrates by chlorination has been abandoned in favor of cyanide treatment by the Crown Reef and the Langlaagte Estate works.

The large profits accruing to some of the mining companies are to a large extent derived from their tailings-treatment. Individual returns show that at several of the leading and most prosperous mines 35 to 50 per cent of their output is due to their using the cyanide process. The Nigel Company's official report for the three months ending September 30, 1893, shows that a net profit of $122,325 was made, of which only $43,970 was derived from amalgamating, and $73,895 from cyanide treatment of tailings and concentrates. The report of the Langlaagte Estate Company for the same period shows a net profit of $51,410, derived from the same source. During 1892, 37,595 oz. were recovered by the process from Robinson tailings, 19,482 oz. from Crown Reef, and nearly 29,000 oz. from Nigel tailings. All companies mentioned use the cyanide process as described in the MacArthur-Forrest patents, and use zinc as bullion precipitant. The amount of gold recovered by the Molloy process (see page 38) is insignificant. The two companies which were described in the official list of the Witwatersrand Chamber of Mines as using that process, are now mentioned as using the MacArthur-Forrest process.

The cyanide plants at Johannesburg are, generally speaking, very similar to each other in their construction, and the description of a typical one, that of the Robinson Company, will sufficiently illustrate their construction. The plant in question, of which the attached plans were designed by Mr. Chas. Butters, consists of twelve circular open leaching vats, each having a capacity of 2,000 cub. ft. and holding 100 tons of tailings and cyanide solution. The vats, built upon elevated arched stone foundations, are filled from a high level tramway above them, and emptied through trap doors in the center into tramcars below.

Next in order is a series of precipitating boxes, designed to continuously precipitate gold from the solution as it passes from the leaching
vat to the sump. The boxes are 20 ft. long, 2 ft. wide, and 2 ft. deep;
they have inclined bottoms. They are divided into compartments a
20 in. in length each. Each chamber contains about 40 lbs. of zinc turnings. Seven compartments in each box are filled with shavings; of
single compartment at the head is left empty to receive any sand that
may be carried through the filter by the solution from the tanks. A
double compartment at the foot is also left empty to allow any gold that
may be carried away by the stream of liquid to deposit before the solution flows into the sump. About 60 tons of solution, which is the quantity required for treating the ordinary daily charge of 225 tons of
tailings, is allowed to run through two zinc boxes in about nine hours.
This solution may carry from 1 oz. to 3 oz. of gold per ton of liquid;
after passing through the zinc boxes it rarely contains more than $2,
and should not contain more than 50 cents if the precipitation has been
properly carried out (Butters and Clennell). Underneath the leaching
vats are four 200-ton sumps, brick and cemented tanks set in the ground.
In these sumps are prepared and stored the solutions for dissolving and
washing out the gold. On the top of them are placed a double set of
duplex pumps, so arranged that both, or either, will throw from any
sump into any leaching vat. On the tram-level above the vat is a 10
horse-power double-drum winding engine and boiler, employed to hoist
the tailings out of the settling pits. The plant includes four furnaces
of large capacity for smelting and refining bullion; also a laboratory
and weighing-room. An elaborate system of tram-lines is laid down, on
both high and low levels, for delivering and discharging the tailings in
the most direct and efficient manner. Works, pits, and dumps are
lighted by electric light. Lathes for turning zinc are employed, and a
12 horse-power engine, with 16 horse-power boiler, supply all power.
The whole is covered by an airy building of wood and iron. The
methods of working are: Hoisting tailings and filling vats; pumping
cyanide solution onto the tailings in the vats to dissolve the gold; running off this gold solution into zinc boxes and precipitating the gold;
return of the cyanide solution into the sump below for repeated use;
collecting, melting, and refining the precipitated gold (M. S. P.).

The report of this company for the year ending December 31, 1893,
shows that 55,200 tons of tailings were treated by the cyanide process,
yielding 17,921 oz. of gold. The cost at the cyanide works is given per
ton of tailings treated, as follows: Wages, 29.62 cts.; supplies, 12.24 cts.;
fuel, 10.48 cts.; cyanide, 48.38 cts.; zinc, 2.26 cts.; filling and discharging vats, 37.62 cts.; royalty, 16.38 cts.; total cost, 173.86 cts. per ton.
The actual cost of treatment per ton, omitting royalty, was 156.98 cts.
The average extraction by the process was 68.7 per cent of the assay-
value of the tailings. An innovation in percolation in the Robinson
Works consists in the circulating system, which has been described by
Butters and Clennell as follows:

"It has been stated that in the usual method of working about a ton
of solution is employed in the treatment of a ton of ore. Since, with
free-milling ore, a much smaller quantity is sufficient to dissolve the
same percentage of gold, it was suggested that the solution from one
tank might be transferred to a second, and be made to dissolve an additional quantity of gold before being passed through the zinc boxes; for

example, it was found at the Robinson Works that 20 tons of solution were amply sufficient to extract 40 oz. of gold from 75 tons of tailings in one tank. It was found that 20 tons of solution sufficed to fill a tank holding the usual charge of 75 tons of tailings, covering the charge to a depth of three or four inches. Instead of replacing these 20 tons of solution by fresh cyanide, the solution filtering through was continually pumped back again into the same tank for about thirty-six hours and then passed through the zinc box. The extraction of gold by this circulation-system was equal to that obtained by the ordinary method, and the consumption of cyanide was much less, since a much smaller quantity of solution was exposed to the action of the zinc. A further modification suggested itself, namely, the transference of the solution charged with gold from one tank to a second and third, in order that it might take up an additional quantity of gold from fresh tailings before passing into the zinc boxes. The advantages of this method are that the solutions from which the precipitate is obtained are much richer in gold, giving a cleaner deposit on the zinc, with much less consumption of cyanide."

In the Durban-Roodeport Company's works the extraction ranged from 67 to 85 per cent. The cost of treating the tailings, including patent-royalty, amounted to $1 54; the profit to $2 87 per ton. During eleven months, in 1893, 79,765 tons of tailings were treated, producing 22,751 oz. of gold, adding about 33 per cent to the total revenue.

The New Chimes Gold Mining Company have commenced to treat their tailings in their own cyanide works only since the beginning of the year; during March, 4,180 tons of tailings realized 709.55 oz. of bullion. The assay-value amounted to (fine gold) 3.33 dwts.; fine gold saved, 2.25 dwts.; extracted, 67.56 per cent; lost in tailings, 32.44 per cent; value of bullion per ton treated, $2 36; expenses, $1 02; profit per ton, $1 32. (G. Halford Smith.)

The financial success of the cyanide process in South Africa is best proved by the dividends paid by the mining companies which use cyanide for tailings treatment. The following is a list of dividends paid during 1893 by companies under such conditions:

Names of Companies.	Dividends.	Amounting to—
City and Suburban	100 per cent	$425,000
Crown Reef	50 per cent	300,000
Durban-Roodeport	45 per cent	282,250
Ferreira	100 per cent	225,000
Langlaagte Estate	30 per cent	705,000
Meyer & Charlton	60 per cent	215,100
New Primrose	40 per cent	391,870
New Rietfontein	25 per cent	200,000
Robinson	8 per cent	1,087,500
Nigel	50 per cent	400,000

The total output of the Rand mines for the year ending June 30, 1893, apart from the cyanide process, was 1,087,058 oz.; by the process this quantity was increased by 226,078 oz., making a total of 1,313,136 oz. From districts not included in the Rand proper, a further recovery of 2,395 oz. was returned, making in all 228,473 oz. due to the working of the process. It will thus be seen that, by the use of the process, the Rand production was increased by 21 per cent. At several of the lead-

5cp

ing and most prosperous mines, 35 to 50 per cent of their gold output is due to the use of cyanide. The report of the Witwatersrand Chamber of Mines gives the output of that district of the Transvaal, for March of this year, at 165,372 oz., from fifty-three mines and three custom works; of which 44,664 oz., of the value of $668,655, were extracted by cyanide from 204,421 tons of tailings, and 1,367 oz., of the value of $20,805, from concentrates by the same means. The returns per ton of tailings averaged 4.37 dwts.; 33.74 per cent of the total month's production of gold is derived from very low-grade material by the process. The cyanide process is the only one which is successfully treating tailings on a commercial scale. Its economical importance for Johannesburg will be evident from the following tables:

Output of Gold in the Witwatersrand District by Mills and Cyanide, in ounces.

Month.	1889. Mills.	1890. Mills.	1890. Cyanide.	1891. Mills.	1891. Cyanide.	1892. Mills.	1892. Cyanide.	1893. Mills.	1893. Cyanide.	1894. Mills.	1894. Cyanide.
January		35,006		52,595	610	72,589	11,971	91,236	17,138	106,263	43,551
February	22,457	36,887		48,532	1,547	75,752	10,897	76,889	16,363	107,829	44,041
March	27,919	37,779		51,348	1,601	81,771	11,473	91,471	20,002	115,882	49,490
April	27,028	38,696		54,726	1,645	82,062	13,500	91,966	20,086		
May	35,028	38,836		53,612	1,061	86,217	13,219	82,906	24,004		
June	30,877	37,419		54,263	1,600	87,752	15,500	96,708	29,199		
July	31,091	39,228	230	52,750	2,174	85,084	16,195	98,278	27,891		
August	30,519	42,807	56	55,524	3,546	83,928	16,392	101,773	34,296		
September	34,143	45,486		61,993	3,609	91,148	16,702	96,322	33,463		
October	32,214	45,248		69,028	3,765	94,836	17,171	101,726	34,956		
November	33,721	46,782		68,020	5,373	89,098	17,696	101,825	36,815		
December	39,050	50,351		71,981	8,331	73,184	17,973	107,060	39,297		
Totals	344,047	494,523	286	694,372	34,862	1,005,421	178,688	1,147,960	330,510		

This table is taken from "Notes on Gold Extraction," by W. R. Feldtmann.

Table showing Companies in the Transvaal Treating Tailings and Concentrates by Cyanide in 1893.

Company.	Tons.	Total Output in ounces for 1893.		
		Plates.	Concentrates.	Tailings.
Champ d'Or	17,896	5,722.7		1,787.17
City and Suburban	49,805	37,777.14		9,034.3
Crown Reef	118,244	51,688.0	723.2	29,679.14
Durban-Roodeport	78,651	37,883.6		22,751.0
Evelyn				2,545.19
Ferreira	47,376	43,978.1		11,697.19
Geldenhuis Main Reef	9,495	3,417.7		372.17
Gipsy	3,931	1,563.6		712.14
Henry Nourse	19,749	15,329.18		6,893.3
J. H. Burg, Pioneer	17,606	9,073.0		1,120.4
Jubilee	43,673	24,774.10	36.0	5,254.16
Langlaagte Estate	222,732	65,812.12	9,047.11	30,050.15
Langlaagte Block B	64,066	19,621.4	100.0	6,869.19
Marais Reef	935	937.5		419.10
May Consolidated	60,298	24,957.4		2,875.0
Meyer & Charlton	34,197	27,328.12		6,854.2
New Heriot	21,455	14,089.2		8,689.18
New Chimes	33,641	14,510.9		7,296.16
New Primrose	141,464	57,574.8		26,203.18
New Rietfontein Estate	24,048	28,168.12	79.18	6,957.15
New Spec Bona	27,289	8,784.5		1,040.0
Nigel	22,273	25,455.0	3,516.2	17,036.8
Orion	34,657	8,318.3		9,677.15
Paarl Central				957.15
Randfontein	54,652	23,310.16		6,623.14
Robinson	94,842	104,222.17	10,659.18	17,921.4
Salisbury	24,786	19,268.18		5,587.6
Simmer & Jack	103,798	38,904.12		767.0
Stanhope	22,858	10,790.8		3,873.16
Treasury	12,429	7,587.12		4,284.16
Village Main Reef	11,607	6,143.17		1,996.8
Vulcan	2,766	764.5		50.0
Wemmer	27,654	22,705.13		3,063.8
Witwatersrand	34,081	12,441.13		7,882.18
Customs Works			38,574.0	35,669.2

Witwatersrand Customs Works, 1893.

African Gold Recovery Company.		Rand Central Ore Red. Company.		Robinson Company.		Total.	
Concentrates.	Tailings.	Concentrates.	Tailings.	Concentrates.	Tailings.	Concentrates.	Tailings.
463.18	4,130.6	9,774.7	31,538.16	28,335.15		38,574.0	35,669.2

Gold is valued at $17 50 for plate gold, and $15 for cyanide gold. (These figures are taken from M. I., vol. 2.)

The table giving the monthly analysis of gold production in the Witwatersrand district, for April, 1894 (see Appendix), which has been published by the Witwatersrand Chamber of Mines, will further illustrate the importance of the cyanide process in that district.

Cape Colony.—In the British Colony at the Cape "the gold mining industry has not developed to such proportions as to lead to the introduction of the cyanide process." (Letter of Secretary of Agriculture, 27th April, 1894.)

B. Australasia.

(a) New Zealand.—A very successful field for the cyanide process has been the eminently progressive British Colony of New Zealand, where various classes of ore, tailings and concentrates, of a very refractory type, have been and are being treated on a large and commercially successful scale. The colony contains the largest cyanide plant outside of South Africa, that of the Waihi Company, with thirteen vats, where ore is treated at the rate of 2,000 tons and tailings at the same rate per month. The Crown mine at Karangahake is equipped with a smaller but equally efficient plant for ore treatment. Smaller plants for treating ores and tailings are distributed over the Hauraki gold fields. An extensive and very successful agitation plant for the treatment of concentrates is connected with the reduction works of the Sylvia Company, Tararu, Thames. The first mine to adopt the process has been the Crown mine in the Upper Thames District. The first plant was erected in an almost inaccessible position in 1889, under conditions which precluded a success. New works have since been erected by Mr. MacConnell, which are in full and successful operation. The ore is clean quartz, with no sulphurets of base metals; the free gold is very finely divided. The silver is in form of sulphide; some of the gold in form of a telluride. The works are described in the New Zealand Government Mining Report of 1893, by Mr. H. A. Gordon, the Government inspecting engineer, as follows :

"The ore, when brought into the works, is first dumped onto a grizzly; what will not go through the bars runs down to the rock-breaker and is broken up to a maximum size of 2 in. diameter, and then falls into the same hopper where the fine material went. The ore passes then into the drying kilns, which are built of brick, the hot air being confined in a long flue, having a series of steps to prevent the ore from traveling down too fast before it gets thoroughly dried. There is a cast-iron plate at the bottom of this flue, which can be turned to allow of the dried ore to pass down into a large hopper, made of steel plates, $\frac{5}{16}$ in. thick, from which the Challenge ore-feeders are fed. These kilns are only for drying the ore, and not in any way to calcine it. There are two of these kilns built on a stone foundation and placed about 6 ft. apart, the foundation going all the way across. The kilns themselves stand about 30 ft. in height, the step-flue being at an angle of about 30° to 40° from the vertical. There is a furnace at the bottom, where either coal or firewood can be used to dry the ore.

"*Stamp Mortars.*—There is first a concrete foundation put in for the stamps, and on the top of the concrete the stamp mortars are placed on the end of a log of kauri, each 18 ft. in length, 4 ft. 8 in. one way, and 2 ft. 2 in. the other. These are firmly embedded in the concrete, and all bolted together, so as to form a solid block of timber standing on end, having a length of 18 ft. 8 in. by a width of 2 ft. 2 in., and on this the four mortars are placed. They are fitted with screens, having the top standing outward at a slight angle, and held to the face of the mortars by means of a long wedge, the gratings being 30-mesh, equal to 900 holes to the square inch.

"*Stamps.*—The stamps are fitted with the latest appliances for raising and holding them up, the cams and tappets being all constructed on the American type. They are intended to make about ninety-two blows per minute, having a drop of six inches. The guides and framing are made of wood. Each ten-head battery is driven by a separate belt, and there is further provision made so that twenty additional stamps can be erected should they at any time be required. The pulverized material from the stamps falls into a chute and is conveyed into another set of hoppers at a lower level than the stamp mortars, and from these hoppers the pulverized dust is taken to the leaching vats.

"*Cyanide Plant.*—This consists of twenty-four wooden vats, each 11 ft. long by 9 ft. wide, and 3 ft. 9 in. deep. In the bottom of these vats there is a false bottom, or grating, placed about 3 in. above the ordinary bottom, and on this false bottom a filter-bed is placed, about 4 in. in thickness, the bottom layer being of coarse quartz-gravel and gradually getting finer up to the top, the last coating being fine sand, having a coarse cloth placed over the top of the filter-bed to prevent the sand from being disturbed, as the vats get cleaned out after every charge of pulverized ore. There are also 14 agitators, 8 of which are 5 ft. deep by 4 ft. 9 in. in diameter, and 6 of them 6 ft. deep and 5 ft. 6 in. in diameter. The agitators and vats are all made of kauri timber, the staves of the agitators being 3 in. in thickness, and the vats being made of partly 3 in. and partly 4 in. timber, and all bolted together. Into each of these vats are placed three pipes, under the false bottom, so that the first, second, and third solutions can be drawn off into separate channels. On one side of each vat there is a door, which can be opened to admit of the material being sluiced out after the whole of the cyanide solution is completely washed out of the ore, the solution passing through a long series of boxes filled with zinc shavings, which precipitate both the gold and silver in the form of a blackish powder. There are also three concrete sumps, each 15 ft. by 12 ft., and 6 ft. deep, capable of holding about 30 tons of the cyanide solution; this is pumped up to the vats on the floor above as required. It is in these concrete sumps where the solution is always made up to the proper strength before being used. It is also proposed to use a vacuum pump to assist the filtration of the solution through the pulverized material in the vats. Annexed are plans of the company's plant, to which the following description or reference applies: At point A, the ore is delivered at the battery and tipped onto grizzly, B; the 'fines' pass through and are conveyed to hopper, D; the 'roughs' pass over the grizzly onto the stone-breaker floor and are passed through stone-breaker, C, and fall into the hopper underneath, marked D; the drying-kiln, E, is charged from this hopper. The ore, after passing through the kiln, being perfectly dry, is run into an iron hopper, G, from where it is automatically fed into stamps, I, by self-feeders, H; the ore, after passing through the stamps, is received in hoppers, J, and then conveyed by means of revolving tube, K, either into trucks for conveying ore to agitation-cylinders for treatment, or, if the ore can be better treated by percolation, to store-hopper, R, in connection with percolation plant, from where it is trucked along the top of and tipped into percolation tanks, S, for treatment. The plant is so arranged that the ore, after it is delivered above the stone-breaker, passes from stage to stage by gravitation, requiring the least possible handling, and thereby reducing the cost of labor to a minimum.

"*Crushing Machinery.*—One Lamberton stone-breaker, capable of reducing 70 tons of ore per day fine enough to feed into stamps; and

20 heads of 9 cwt. dry stamps, crushing 30 tons of ore per day through a 30-mesh screen.

"*The Percolation Plant* consists of 24 tanks, capable of holding each a charge of 7 tons of finely pulverized ore. The bottom of each tank is covered with a sand and gravel filter. The ore is trucked into the tanks from the storage hoppers. A dilute solution of cyanide is then run on the top, and allowed or assisted to percolate through the body of the ore. As the solution percolates, it is carried away from underneath the filters by means of iron pipes, and permitted to run through a series of boxes of zinc turnings.

"*Agitation Plant.*—This consists of sixteen wooden tubs, fitted with revolving paddles, in which the ore and cyanide solution are agitated together until the gold and silver are dissolved. The pulp is then filtered by means of filter-presses, and the bullion deposited from the solution on the zinc, as already described. The extraction of bullion is given as 93 per cent of the gold and 79 per cent of the silver assay-value. The cost of treatment is $3 50 per ton." This is the only company in New Zealand which does not pay any royalty to the owners of the Mac-Arthur-Forrest patents, the patentees owning part of the mine. The total bullion-value produced by the cyanide process in these works amounts to upward of $142,000.

Another mine of importance, the Waihi Company, has recently adopted cyanide treatment for their ores, supplanting unsatisfactory pan-amalgamation. The ore of that mine is very similar to that of the Crown mines. The bullion recovered by amalgamation has never exceeded 66 per cent of the gold and 40 per cent of the silver assay-value. Experiments on a large scale, made nine months ago, led to the construction of an extensive percolation plant, by which upwards of 20,000 tons of ore have been already successfully treated. The extraction varies from 89 to 91.8 per cent of the gold, the silver extraction from 46.5 to 51 per cent of assay-value. The cost per ton for cyanide and zinc is $1 37½. The gold returns from cyanide treatment are 25 per cent higher than from pan-amalgamation. The ore, which required 60-mesh screens for amalgamation, is sufficiently fine for cyanide if passed through 40-mesh, which means an increased output from the mill of at least 25 per cent, the running expenses remaining virtually the same. Eventually 30-mesh wire gauze may be used. The strength of solution used is from 0.25 to 0.4 per cent. The percolation and subsequent washings can be done in four days. No difficulties have been found in percolation, as the dry ore does not form slimes as wet ore probably would. After the first percolation is finished, the subsequent washings are hastened by atmospheric pressure by means of a vacuum pump. The extra profit by cyanide treatment of Waihi ores over pan-amalgamation amounts to about $3 75 per ton. The company has been experimenting with the Otis crusher, as a substitute for the dry-crushing stamp battery; the results have been unsatisfactory. A royalty of 7½ per cent on the bullion produced is paid to the owners of the MacArthur-Forrest patents, the Cassel Company of Glasgow. For the information in reference to the Waihi Company's cyanide operations, I am indebted to the company's manager, Mr. R. Rose.

A fuller description of the working of the process has since been given by Mr. Barry in the report of Mr. H. A. Gordon, the Inspecting Engineer of Mines to the New Zealand Government (New Zealand Mining Report, 1894), as follows:

"The ore is first dried in open kilns, excavated in tufaceous sand-stone. These are 37 ft. deep by 20 ft. in diameter at the top, and taper down to the bottom, where they are finished off with a brick arch, having a door and an iron chute for discharging the dried ore into trucks. These kilns are first charged with wood and ore in layers, each layer of wood being about 5 ft. apart. After the kiln is fully charged, the wood is lighted, and after being all burned up, about one half of the charge is withdrawn—50 tons—and another 50 tons of raw ore, together with wood, added on the top; after which about 50 tons is withdrawn every third day. This method of drying the ore is found to be very economical as regards fuel, as there is not a large surface of cold material to heat up, as is the case with smaller kilns, which are emptied at each charge. The cost of firewood used in large kilns is $37\frac{1}{2}$ cents per ton of ore dried. After the ore is taken from the kilns, it is then put through the rock-breaker, from which it falls into a hopper, and thence, by automatic feeders, it is fed into the stamp-mortars, when it is pulverized until it passes through a 30-mesh and sometimes a 60-mesh screen. It is intended in the future to use a 40-mesh standard. As the pulverized dust passes through the screens it falls into a narrow trough, when it is conveyed by means of an Archimedian screw into a dust-bin at one end of the battery, and from this bin the pulverized material is lifted with a bucket-belt elevator and discharged onto an 8 in. rubber belt with rope edges, and conveyed to and across the hopper 110 ft. long, running the entire length of the cyanide plant-house. This hopper has twenty doors for discharging the sand into the trucks, which are then run straight out over the percolating vats into travelers, running on rails, which are fitted with hand-traversing gearing, enabling a truck to be tipped at any part of the vat. This is an important point, as sand has a tendency to pack if moved about or touched in any way after being tipped into the vat. As a further preventive against packing, there is a small traveler running under the main traveler, with a platform just at the height that the sand is to be filled up to. All trucks are tipped over this plat-form, which breaks the fall and throws the sand off in a light shower all round. When the vats are filled up to a depth of about 2 ft., a strong solution of cyanide—0.4 per cent—is introduced into the bottom of the vat under the filter-cloth, and forced up through the sand until it stands about 2 in. above it. The solution remaining under the filter-cloth is then drawn off, and filtration commences; the 2 in. on the surface taking about twenty-four hours to percolate through. After the whole of the strong solution has been taken out of the ore, a weak stock solution is run on the top of the ore to a depth of about $6\frac{1}{2}$ in. The cock connecting with the vacuum cylinder is then opened, and in about thirty hours the second solution has passed through; after which about 10 in. of water is run onto the top, and when this has gone through the ore the operation is completed. The sludge-door in the vat is opened, and the sand sluiced out by means of two 2 in. hose-pipes under a head of 150 ft. The vats are all circular, 22 ft. 6 in. in diameter, and 4 ft. in depth, of which 5 in. is taken up by the filter bottom, which consists of a wooden grating with edges rounded off on the upper side, having a strong Hessian cloth laid over the top, which acts as a filter. The vats are made of kauri timber, 3 in. in thickness; the bottom is held together by six bolts of $\frac{3}{4}$ in. diameter. The staves are about 3 in. in width, joined close, having the bottom rebated into the sides. Each vat is held together by five round-iron hoops, three of which are $\frac{3}{8}$ in. and two 1 in. in diameter, having three turn-buckles on each hoop. The plant consists of thirteen of these vats and two sumps of the same diameter, but 6 ft. in depth. Each vat holds 30 tons of ore for treatment; and it takes about four days to fill a vat, treat the ore, and have it ready for filling again. The precipitation boxes are 16 ft. long, 2 ft. deep, and 17 in. wide, divided into twelve divisions, of which the first and last are sand filters, to clean the solution going in, and to prevent any gold slimes from being washed out.

"The cost of treatment is being reduced every month; at the present time it amounts to about $3 25 per ton. This includes drying, milling, treatment by cyanide, and all expenses, except the royalty paid to the owners of the MacArthur-Forrest patents, from the time it leaves the mine-hopper until the bullion is in bars."

The tailings from the former pan-amalgamation process are being worked in a cyanide plant erected for that purpose by the Cassel Gold Extraction Company. These works were completed about the end of February, 1894. Mr. H. A. Gordon gives the following description, illustrated by plans which I reproduce: "The works are situated in a hollow below the tailings dam, so as to allow the tailings to be run at a good grade into the percolation vats, and from there to be discharged by sluicing, without the necessity for any lifting or rehandling.

"The building has a frontage of $116\frac{1}{2}$ ft., and is 77 ft. in breadth, and includes laboratories and offices, situated in a 'lean-to' at one end and communicating with the main building. The plant consists of eight circular percolation vats 20 ft. in diameter and 4 ft. in depth (internal measurement), arranged in two rows, and having an intermediate discharge-launder, toward which the vats have a slope of 2 in. to facilitate the flow of solutions and the sluicing-out of residues. All the vats are built of specially selected and well seasoned heart-of-kauri, the timbers being 3 in. thick. The sides are hooped with $1\frac{1}{8}$ in. iron bolts, connected and screwed up by nuts and cast-iron boxes, there being three boxes to each ring. The bottom planks are bolted and dowelled tightly together independently of the sides. The filters at the bottom consist of a foundation of 2 by 3 in. slats, 9 in. apart, covered by 1 in. molding, which supports the canvas strainer. This filter is very easily laid, and is most effective in practice. Each vat is provided with a cast-iron door 18 by 12 in., fixed at the bottom of the side near the discharge-launder, for the sluicing of residues. There are two sumps of same size and design as the percolators, and situated between the percolators and front of the building, and on a sufficiently low elevation. The sumps are floored over. In the same line are placed the reservoir and cylindrical vacuum chamber, 13 ft. by 3 ft. 9 in., under which latter is provided a small rectangular tank, 12 ft. by 8 ft. by 18 in. deep, capable of holding contents of vacuum chamber. The reservoir is 13 ft. 9 in. diameter by 5 ft. deep (inside measurement), and is at such an elevation as to permit solutions to flow therefrom into percolators. There are three extractor boxes, 12 ft. 8 in. by 19 in., with side discharge for slimes and a settler for cleaning-up. The tank for dissolving the cyanide is an iron pan about 3 ft. 6 in. in diameter by 2 ft. 6 in. in height, and is capable of dissolving four boxes—i. e., 1,000 pounds—of cyanide per

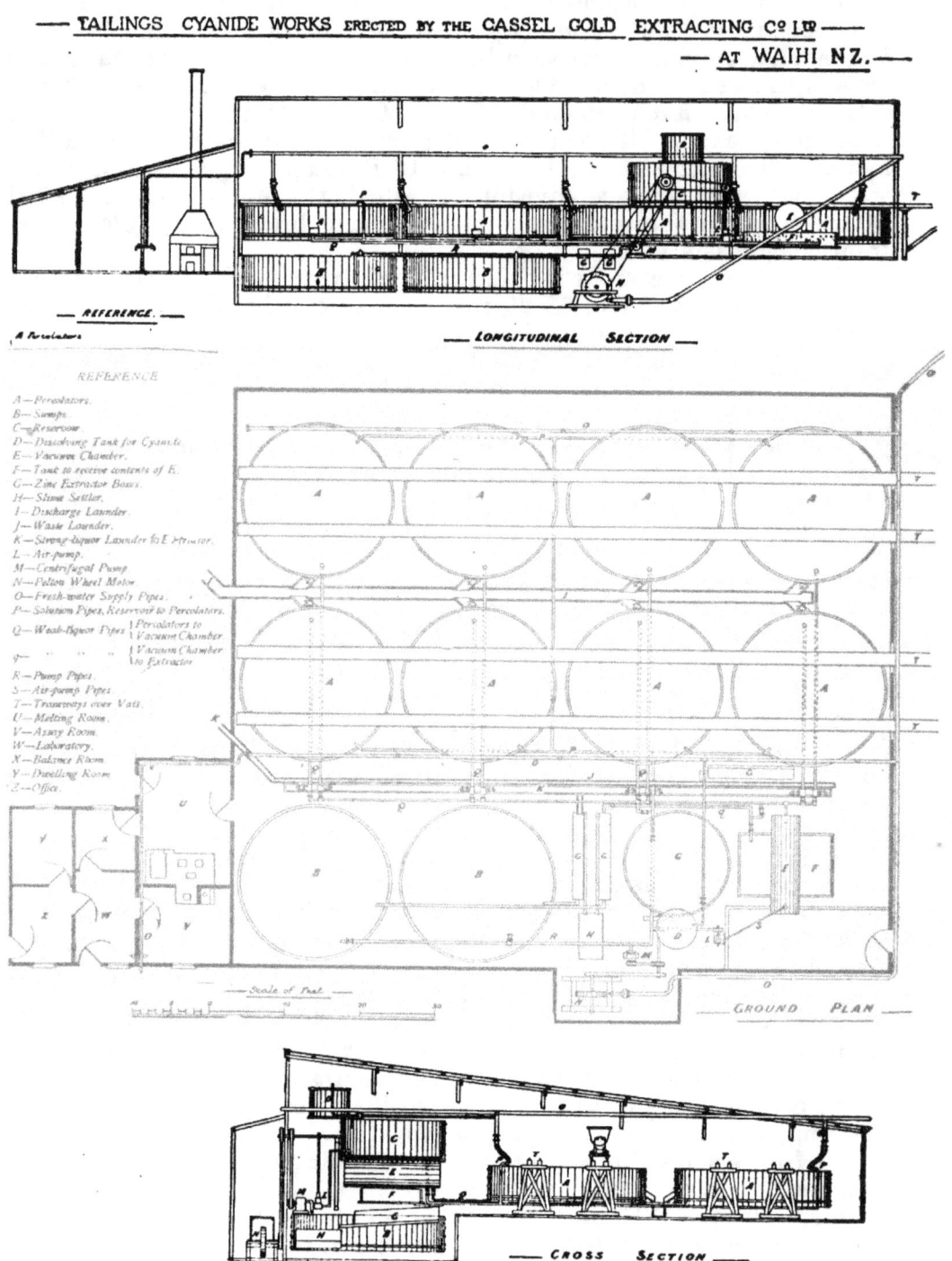

TAILINGS CYANIDE WORKS ERECTED BY THE CASSEL GOLD EXTRACTING Cº Lᵀᴰ AT WAIHI N.Z.

LONGITUDINAL SECTION

REFERENCE

A—Percolators.
B—Sumps.
C—Reservoir.
D—Dissolving Tank for Cyanide.
E—Vacuum Chamber.
F—Tank to receive contents of E.
G—Zinc Extractor Boxes.
H—Slime Settler.
I—Discharge Launder.
J—Waste Launder.
K—Strong-liquor Launder to E Extractor.
L—Air-pump.
M—Centrifugal Pump.
N—Pelton Wheel Motor.
O—Fresh-water Supply Pipes.
P—Solution Pipes, Reservoir to Percolators.
Q—Wash-liquor Pipes { Percolators to Vacuum Chamber
q— { Vacuum Chamber to Extractor
R—Pump Pipes.
S—Air-pump Pipes.
T—Tramways over Vats.
U—Melting Room.
V—Assay Room.
W—Laboratory.
X—Balance Room.
Y—Dwelling Room.
Z—Office.

GROUND PLAN

Scale of Feet

CROSS SECTION

day. It is so arranged that the requisite amount of strong solution may be run into the reservoir by simply turning a handle. A 4 in. centrifugal pump serves for returning the solution from the sump to the reservoir, and also an 8 in. vacuum pump, which is capable of producing a vacuum of 26 in. of mercury. A line of pipe runs along above each row of percolation vats, with a connection at each tank for the hose and nozzle. One man can empty a vat containing upward of 40 tons in two hours. A tramway connects the tailings-dams with the works, and two sets of lines run over the top of each vat, so that the tailings may be equally distributed without the necessity for handling. The chief characteristic of the plant is its extreme simplicity and the easy access to any portion of it; the absence of any subdivisions or partitions within the main building exposes the whole of the plant constantly to the eye of the operator.

"The system employed of running the solutions into parallel launders instead of pipes, enables the solutions from each vat to be separately and readily sampled and any mishap may be at once detected. The usual method of procedure is as follows: Side-tipping trucks are run from the tailings-pit over the top of the vat which is to be charged. The contents of the trucks are tipped onto cross-bearers resting on struts, which serve to break the fall of the tailings, and to divide them equally over the bottom of the vat. Both tramway and bearers are supported entirely independent of the vats, so that no vibrations may be communicated to the latter. A charge consists of sixty-five truck-loads—about 33 tons, dry weight—and as soon as the vat is full, 'strong' solution—about 6 tons of 0.7 per cent—is run onto the top from the elevated reservoir. Provision is made for either upward or downward percolation, but the latter is usually adopted. The solution is now permitted to gravitate through the mass of the charge, and to eventually percolate through the false bottoms into the series of launders in which it is conducted to No. 1, No. 2, or No. 3 zinc precipitation box, according to its strength in bullion and cyanide. About twenty-four hours after the 'strong' solution, about an equal amount of 'weak' solution (0.25 per cent) from the sumps is pumped on and allowed to gravitate. The residues are now washed with about 10 tons of water in two charges, which are rapidly drawn off by suction, and which displace the 'weak' solution and leave the residues free of either dissolved bullion or cyanide. The solutions run from the zinc boxes to the sumps, whence they are pumped to the reservoir or percolation vats, to be used over again for sluicing or weak solutions as required. A clean-up of the gold in the zinc boxes takes place fortnightly."

The New Zealand Mining Report of 1893 contains the description of an interesting experiment which was made some time ago in the Waihi works, with an apparatus by which ore was intended to be rapidly extracted by a cyanide solution acting in a jigging motion on the ore in iron cylinders. The cylinders used were, however, too long and narrow, containing as they did some 10 ft. in depth of ore, which the solution had to be forced through. The effect of this was that the solution could not be made to percolate through the whole of the ore, but passed up between the cylinder and the ore, the solution being forced into the cylinder by a pump, at a pressure of 100 lbs. per square inch. This pressure should have been sufficient to force the solution through, but as the pulverized material offered greater resistance than the contact between the material and the side of the cylinder, the solution went through the weakest spot, and had little effect on the ore. The process ("the Bohm Process") proved a failure.

The unquestionable success of the percolation process with the Crown and Waihi mines has led to its adoption by a number of other companies which treat either ores or tailings by the process, as the Te Komata and Waiorongomai mines at the Upper Thames, and two or three companies on the Kuaotunu gold field. The Kuaotunu ore, in which the gold is exceedingly fine, is especially adapted for the treatment, the only difficulty experienced—a mechanical one—is caused by the amount of slimes formed by some of the ores, which interfere with filtration. The plants on that field do not offer any special point of interest; they are of small extent and give satisfactory results. The best one, that of the Tryfluke Company, will, however, be described on account of the attempts made therewith to run the tailings direct from the battery into the percolation vats. The plant consists of four working tanks, each 12 ft. wide, 16 ft. long, and 4 ft. deep, having a filter-bed of 3 in. in thickness, covered with a coarse cloth. The depth of ore in the tanks is about 3 ft. 6 in. and about 6 tons of 0.25 per cent cyanide solution are used per charge. This is what is termed the strong solution. The tap, which allows the filtrate to flow away, is so regulated as to take about 24 hours for that purpose. After flowing through six compartments of a filter box, which are filled with fine zinc-turnings, the solution passes into a sump, 18 ft. long by 14 ft. wide and 4 ft. deep, from where it is pumped into the tank again, thus forming the second solution. This latter is allowed to filter through as fast as possible, and, after going through the boxes filled with zinc, it flows into another sump of the same dimensions as that already mentioned. The ore is then washed with pure water, after which the material is shoveled out of the tanks and run onto the waste dump. The second solution in the sump, previously referred to, is pumped into a reservoir placed at a higher level than the working tanks. This reservoir is 10 ft. long, 8 ft. wide, and 5 ft. deep. The solution is made up to the required strength before again being used. The company tried to run the tailings directly into the tanks from the battery, but they, like others, found that the amount of slimes in the ore prevented the cyanide solution from filtering, and they are now making arrangements to run the tailings into a large pit, from which they will be lifted into the cyanide tanks. (Extract from N. Z. Gov. Mg. Rep.)

All works so far referred to are situated in the North Island; on the large gold fields of the South Island the process has not found more than experimental application. Experiments have been made with gold-bearing cement from the extensive deposits on the west coast, where almost inexhaustible quantities of conglomerates, containing black magnetic oxide of iron and very small quantities of gold, are found. Such experiments were made, for instance, in the Reefton School of Mines, by treating the cement in lumps, but they were not always successful, apparently for mechanical reasons. When the cement is crushed, a very good percentage is said to be extracted, the gold being fine and well suited for the purpose. Tests with tailings from the Inangahua River gold fields have given good results, and a plant for working a considerable deposit of tailings is nearing completion at Boatman's.

The only instance in New Zealand where the agitation system has

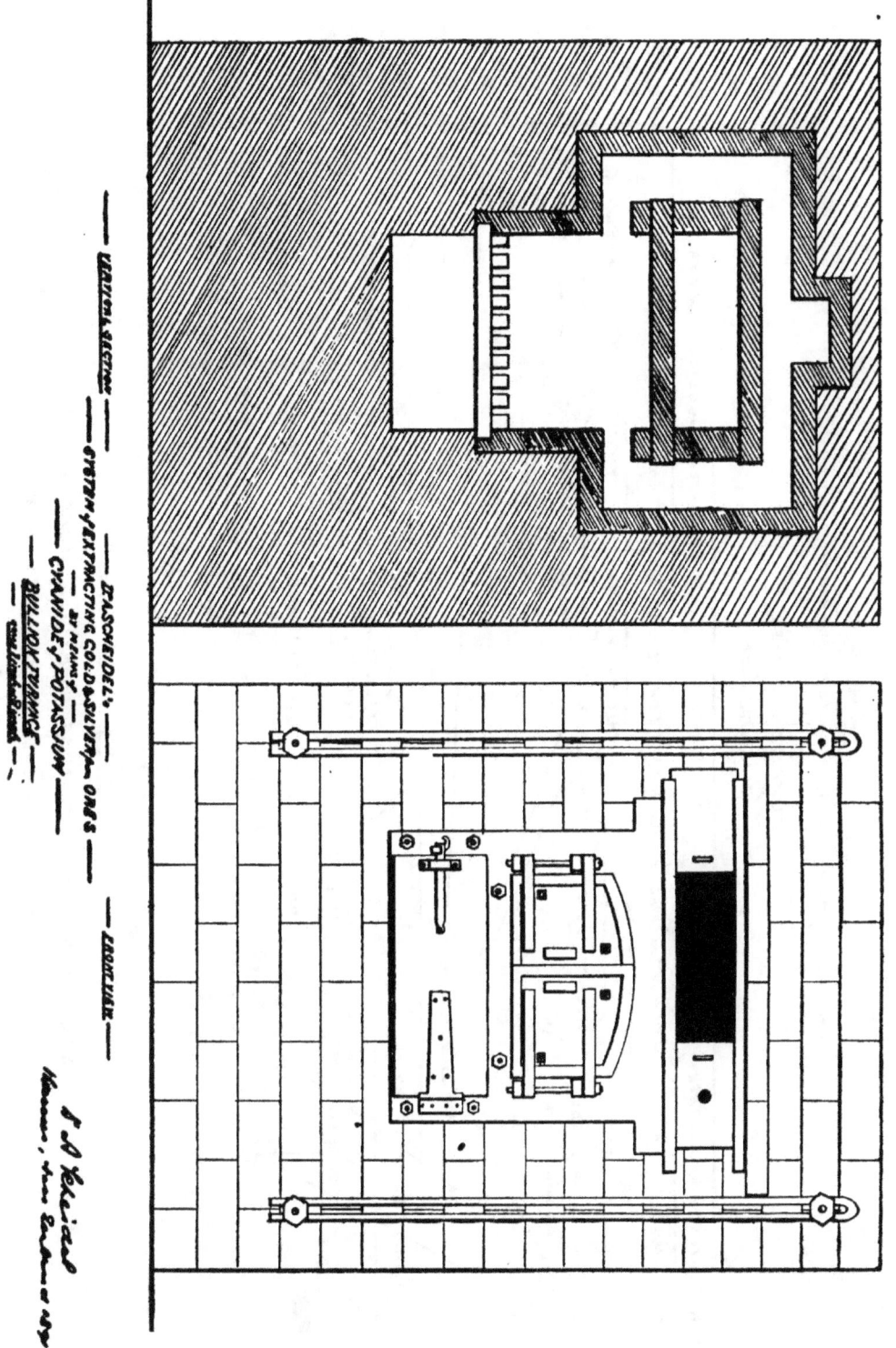

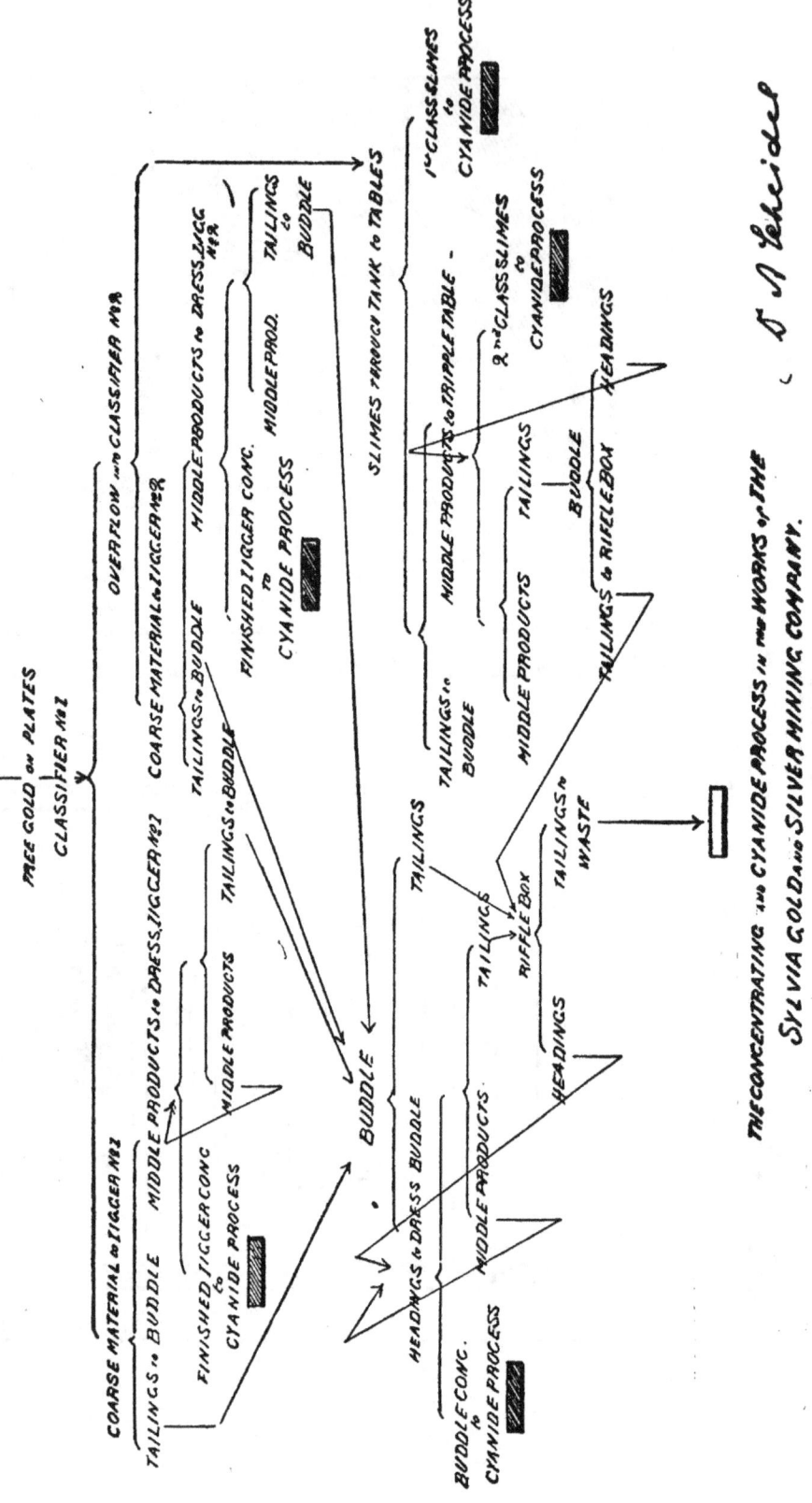

THE CONCENTRATING and CYANIDE PROCESS in the WORKS of THE
SYLVIA GOLD and SILVER MINING COMPANY.
TARARU, NEW ZEALAND.

been worked on a large scale is in the works of the Sylvia Company in Tararu, Thames, where concentrates of a very complex character have been treated with full success by the author. The ore of the mine named contains, in the deeper levels, a high percentage of galena, zinc-blende, and copper and iron pyrites. Most of the bullion is contained in the sulphurets and cannot be saved by amalgamation, nor is there any opportunity for smelting the sulphurets after concentration or for remunerative exportation of the same. After successful trials with the concentrates subjected to cyanide treatment, I constructed an agitation plant, of which I append my plans, reproduced from the N. Z. Gov. Mg. Rep. 1892. The concentrates in question are classed as jigger concentrates, first-class slimes, second-class slimes, and buddle concentrates. They are named in the order as they are obtained during the dressing process. The finest products contain the most galena; they are the richest in gold and silver, and give the highest percentage of extraction. A detailed description of plant and extraction process is given in the mining report above referred to. I give here only the most salient points: The plant, constructed of wood (kauri pine), consists of three agitators, 6 ft. in diameter by 6 ft. in depth; three Scheidel's vacuum filters (patented) of 35 ft. square filter surface each; the necessary solution tanks, pipes, and pumps, and bullion roasting and melting furnaces. The experience gained with this plant, which works well with the exception of the faults which necessarily adhere to the use of wood for cyanide plants, induced me to construct, later on, the plant for the Utica Mine, California, of steel. This plant, being free from the only fault, that above mentioned, answers the purpose to perfection. The results of extraction of the Sylvia concentrates vary in accordance with the quality of the material, the slimes giving generally better results than the coarser products, and the richer first-class slimes return a higher percentage of gold and silver than the second-class slimes, which are of lower grade. Eminently satisfactory results have been obtained from the best slimes, from which as high as 96.45 per cent of the gold and 94.59 per cent of the silver assay-value have been extracted. The average extraction of 100 tons amounted to 86.11 per cent of the gold and 67 per cent of the silver; corresponding with 85.22 per cent of the total value. The extraction of the least suited material, the jigger concentrates, coarse and low grade, amounted to 80.32 per cent of the gold and 50 per cent of the silver. The average extraction of all classes of concentrates amounted to 82.67 per cent of the assay-value. The total amount of bullion extracted by me from about 300 tons of material amounted to upward of $51,000. The time of agitation and the strength of solution vary in accordance with the quality of the material. The quantity of cyanide used for the highest grade ore amounted to less than 1 per cent, and for low-grade material considerably below 0.5 per cent of the ore. The time of agitation varied between six and twenty-four hours. The method of working the plant, which permits the treatment of twenty tons per twenty-four hours, is identical with that of the Utica plant, which I propose to describe in detail. The Sylvia Company enjoyed, on account of the plant being the pioneer plant of its kind, special privileges in reference to royalty for the use of the reagent.

New Zealand is among the few countries, outside of Africa, where cyanide treatment of ores, tailings, and complex concentrates has been in all instances a perfect success. Many of the ores of the Colony are particularly suited to the treatment. Its still more extensive application is prevented by the royalty charged by the owners of the MacArthur-Forrest patents, in reference to which fact I quote the Government Inspecting Engineer of the Colony, Mr. H. A. Gordon, who says (Gov. Min. Rep. 1893):

"There is no gainsaying the fact that the cyanide of potassium is a good reagent for gold, and undoubtedly the best agent for extracting gold, especially where the latter is in a very finely divided state among the ore; but the royalty charged is prohibitive, and it will never be largely used until it is lowered. If the Cassel Company (the owners of the MacArthur-Forrest patents) would have been content with about 2 per cent royalty, the process would have been almost universally adopted by every company in New Zealand."

The following particulars represent the results of the working of the cyanide process on the New Zealand gold fields up to the end of December, 1893:

Names of Companies.	Bullion, in oz.	Bullion, value.
New Zealand Crown Mines, Karangahake, from ore	15,064	$145,930
Sylvia Gold Mg. Co., Thames, from concentrates	14,552	51,965
Tryfluke Gold Mg. Co., Kuaotunu, from tailings	3,072	35,895
Great Mercur Gold Mg. Co., Kuaotunu, from tailings	605	6,410
Te Aroha Gold Mg. Syndicate, Waiorongomai, from ore and tailings	1,981	4,595
Te Komata Gold Mg. Co., Upper Thames, from tailings	1,741	5,155
Red Mercury Gold Mg. Co., Kuaotunu, from tailings	149	1,230
Silverton Gold Mg. Co., Waihi, from tailings	299	2,105
Waihi Gold Mg. Co., Waihi, from ore	1,097	9,045
Welcome Gold Mg. Tailings, Boatman's, from tailings	200	2,255
Totals	38,760	$264,585

The returns by cyanide during the twelve months ending March 31st last are: Bullion obtained from ore, 14,774 oz.; from tailings, 12,478 oz. The year 1894 promises to be even a more successful one, six new cyanide plants being in course of erection. The bullion obtained by cyanide during the quarter ending 30th June amounted to 13,030 oz. from ore, and to 7,073 oz. from tailings. Fifty-two per cent of the total bullion product in the North Island of New Zealand was produced during that period by the cyanide process.

In the other colonies of the Australasian group the process is comparatively slow in being introduced. The monetary crisis that has prevailed in the colonies, combined with prejudice and skepticism on the part of the mining community, have retarded its introduction.

(b) **Tasmania.**—In this colony the process has not yet been introduced on a working scale. So far, the only use that has been made of it, has been on a small and experimental scale, and in a very imperfect manner. (Letter from Government Secretary for Mines, January 15, 1894.)

(c) **Western Australia.**—Here the process has not been adopted on a large scale, "the mining industry being yet in its infancy. The process is, however, considered the best known for Western Australian gold ores, and some minor experiments have been made with tailings, giving good results." (Letter from Government Secretary of Mines, January 19, 1894.)

(d) **South Australia.**—This colony has several plants for the working of the cyanide process, one of which, that at Mt. Torrens, is being worked as a custom works by the Mines Department of the State for the purpose of giving the mine owners an opportunity to have their ores tested. A Government plant undoubtedly inspires miners and prospectors with confidence. A charge just sufficient to cover cost of treatment is made by the department.

"The plant has now been in operation for some weeks, and the treatment of an ore parcel from the Blacksnake Mine has been completed. The ore contained on an average 16 dwts. 15 grs. of gold; of this, 10 dwts. 8 grs. were saved by the battery, 4 dwts. 9 grs. by cyanide, leaving 1 dwt. and 22 grs. in the tailings. The ore contained quartz, hematite, and about 4 per cent of iron pyrites." (Letter from Secretary for Crown Lands, May 23, 1894.)

A plant for treating 500 tons per month is at work on tailings at the Virginia Gold Mining Company's property, and is doing good work. The tailings from the battery are allowed to dry, and are then trucked into vats of a capacity of between 25 to 30 tons, 16 ft. in diameter by 5 ft. in depth. The sumps are built of cement, 16 ft. by 14 ft. by 12 ft. The consumption of cyanide is about $1\frac{1}{4}$ lbs. per ton of tailings. The tailings before treatment assay from 10 to 15 dwts., after treatment, 1 dwt. 7 grs. per ton. The bullion is refined with nitre. The vats are charged twice a week. (Government letter.)

(e) **Queensland.**—At Charters Towers, a plant capable of treating 800 tons per month has been erected by the Australian Gold Recovery Company, Lim., the owners of the MacArthur-Forrest patents for Australia, from which the following information has been obtained: It is a custom plant. The chief material treated is sludges, which are purchased from the surrounding mills in varying quantities. These sludges are concentrates which have been submitted to fine grinding and amalgamating in berdans; this material is, if necessary, mixed with coarse sand or tailings, and treated by percolation; vacuum pumps are used to assist. Different classes of ore of a refractory character are treated in the works, and the conditions of treatment are varied with the character of the ore. Operations were started in August, 1892, since which date about 9,200 tons of sludges have been treated, with a return of 9,633 oz. of gold. "At Croydon a tailings plant has been erected at the Cumberland property of a capacity of 1,500 tons per month. Plants for the treatment of 2,000 and 1,000 tons per month are in course of erection for the Croydon Quartz Crushing Company and the Pioneer Gold Mining Company, respectively." (Australian Gold Recovery Company.)

(f) **New South Wales.**—In this colony the cyanide process is at work only at the Mitchell's Creek gold mine, where the plant and operations are described by Mr. James Taylor, the Government metallurgist, as follows: "The works have been erected for the treatment of a dump of old tailings estimated to contain about 18,000 tons, and found by careful sampling to contain 8 dwts. 4 grs. of gold and 11 dwts. 10 grs. of silver per ton. The plant consists of two 400-gallon iron tanks, two storage vats, six percolating vats, two sets of ten precipitating boxes, two sumps, iron pipes for conveying the solu-

6CP

tions and washes, side-tipping wagons and tram-rails for charging and discharging the percolating vats, steam boiler and pump to return the cyanide solution from the sump to the storage vat, small muffle furnace for roasting the precipitated gold with its admixture of zinc and copper, laboratory and assay furnaces.

"In the two iron tanks the cyanide stock solution is prepared by dissolving crude potassium cyanide, suspended in a wire gauze tray in water. The crude salt contains about 75 per cent of pure potassium cyanide, and the stock solution is made up to a strength of from 10 to 25 per cent. From the iron tanks the solution is run through a canvas-bottomed box, which serves as a filter, into one or the other of two storage vats, as it is needed to bring up the strength of the returned liquor to the standard required for the treatment of the next charge of ore, say about 0.7 per cent of cyanide. These two wooden storage vats are 16 ft. in diameter and 5 ft. deep, and are placed sufficiently high to discharge into the next series of vats by means of iron pipes. The six large percolating vats, each 18 ft. in diameter and 5 ft. deep, are provided with filter-bottoms, built up by laying ribs of wood, notched on the under side, along the bottom of the vat at regular intervals. On these ribs is spread cocoanut matting, and over this comes a layer of woolpack. The edges of the filtering cloths are well caulked along the sides of the vat. The vats are charged with tailings by means of side-tipping wagons, carried on an over-head tram-line, and the exhausted tailings are discharged by being shoveled through an 18 in. hole in the side of the vat, near the bottom, into a wagon running on rails and leading to the waste dump. Each percolation vat receives a charge of 35 tons of tailings, and two vats are emptied and refilled daily during the day shift, so that 70 tons of tailings are treated every twenty-four hours. The fresh charge of tailings is first soaked with returned solution, which begins to make its appearance at the bottom of the vat in about three hours from the time of its application. A solution of 0.7 per cent is then turned on, and this is allowed to act for about twenty-five to thirty hours; it is then drawn off, running direct to the precipitating boxes. About twenty inches of liquid from the storage vat is sufficient to soak the charge, and a similar amount of the reinforced solution is enough for the gold extraction, after the application of which water is run on to wash out the cyanide solution. Each of the percolation vats can be supplied with cyanide solution from the storage vats by means of a 2½ in. pipe, and with water through a 2 in. pipe, for washing after the cyanide solution has been drained off, or before the cyanide solution has been added should that be found necessary, as is sometimes the case. The tailings are under treatment in the vat during a period of about sixty-four hours from the completion of charging to the commencement of discharging the vats.

"An inch and a half pipe proceeding from the bottom of the percolating vats takes the gold solution to the precipitating boxes, where the liquid is caused to pass upward successively through a series of ten boxes, filled with fine zinc turnings, made on the spot from the zinc linings of the boxes in which the potassium cyanide has been imported. The gold and silver are precipitated upon the zinc as a fine black slime, accompanied by any copper that may have been taken up by the solution from the tailings. After passing these boxes, the solution, almost or altogether free from gold, is collected in a couple of sumps, each 16 ft. in diameter and 5 ft. deep, from which it is pumped back to the storage vats, ready for another application, either for soaking, or, when suitably reinforced, for gold extraction.   When a clean-up is being made, the zinc in the boxes is well stirred to shake off the slimes, which are then washed out through a plug-hole into a launder, where they are collected, and either roasted and melted with a little flux, or treated with acid and then melted; the mode of procedure depends upon the amount of base metal mined with the gold and silver.   All solutions going from the per-colating vats pass through the precipitating boxes, excepting of course the wash water retained as moisture in the tailings when the vat is emp-tied.   The tailings contain, as noted, 8 dwts. of gold and 11 dwts. 10 grs. of silver to the ton.   The returns for a recent run of ten weeks show that 64 per cent of the gold and about the same of silver was actually recovered; but by assay of the tailings after treatment it appears that 70 per cent of the gold had been removed; hence something like 6 per cent seems to be locked up in the plant, and may be obtained later. The cost of cyanide during the same period of ten weeks amounted to $2 58 per ton of tailings, and the total cost of treatment to $3 38 per ton.   The tailings contain from 0.25 per cent to 0.50 per cent of copper, the presence of which increases the amount of cyanide used by from 75 cts. to $1 per ton, and further acts injuriously, as it is precipitated by the zinc with the gold and silver and debases the bullion obtained, thus necessitating refining operations.   The force employed in the works consists of one scientific manager, two men on alternate twelve-hour shifts to attend to the circulation of the solutions, nine men to fill and empty the vats, one man to prepare the cyanide solution and to do odd jobs, and a laboratory boy.   The cost of the plant was approximately $10,000; it was erected under the auspices of the Australian Gold Recov-ery Company, Limited" (MacArthur-Forrest patents).   During six months 9,972 tons have been treated, with an extraction of 2,512 oz. of gold.

(g) **Victoria.**—"A plant of 2,000 tons per month is being constructed at the New Golden Mountain Gold Mining Company's property, and it is proposed to treat the ore at these works directly by cyanide, without any previous treatment or battery amalgamation". (The Australian Gold Recovery Company.)

C.   The United States of America.

The introduction of the cyanide process in the United States as a metallurgical process on a commercial basis has so far been slow; the ter-ritory is vast, the mining districts are widely scattered, and there is an almost invincible prejudice against any new process, particularly patent processes, owing to the innumerable failures in which for years past much money has been lost.   Of late the process has been tested on many ores from almost all mining camps in the United States.   It has been found successful, on a small scale, in many instances.   Its tech-nical application has, however, not always been the expected success, for which fact various causes are responsible.   In many instances plants have been erected where either the ore is unsuited for the process or where the supply of suitable ore is insufficient; in other instances the working of the process has been intrusted to incompetent hands, which

naturally led to a failure. The MacArthur-Forrest patents and the
Simpson patent are owned in the United States by the Gold and Silver
Extraction Company of America, Lim., of Denver. A number of extraction works have been erected all over the United States by mining companies and other parties. To obtain reliable information of their results
and their plants has proved possible only in a few instances, and it is
hardly possible to form from such information an adequate idea about
the success of the process in this country.

(a) **Utah.**—The first mill to operate cyanide treatment in this Territory is owned by the Mercur Gold Mining and Milling Company. That
company had just completed a pan-amalgamation plant at a cost of
$30,000, which proved a failure, only 20 per cent of the gold being
recovered, when small tests with cyanide, followed by good results, led
to the introduction of the process on a large scale, which has since proved
a full success. For the following description, by Louis Janin, Jr., I am
indebted to the Engineering and Mining Journal (Oct. 7, 1893):
"The ore passes through a Dodge rock-breaker, and is crushed by two
sets of Wall's corrugated rolls; the first are set to one half inch, the
second to one quarter inch. This very coarse ore (over 20 per cent of
the product which goes to the leaching vats does not pass a half-inch
mesh) is treated by cyanide percolation. The dimensions of the vats
are 12 ft. 8 in. diameter; depth to false bottom, 35 in.; giving a capacity
of about fourteen tons when the vats are filled to within 6 in. of the
top. In consequence of the crudeness of the crushing, the time of
leaching varies greatly; it occupies between ten and two hundred
and forty hours. It is claimed that the ore (which is a silicious sur-
face-ore with the gold finely divided) is singularly constant in value
and quality; the wide differences in the time of treatment are ascribed
to the differences in mechanical condition. As a rule, the cyanide solution is left standing with the ore for twelve hours; it is then passed
through continuously, until practically all gold is extracted; the time
required varies from thirty-six to forty-eight hours. The percolation
liquor passes through zinc boxes 40 ft. long, and is returned to the stock
solution tank, where its proper strength is made up again by addition of
cyanide. After the ore in the tank has been leached sufficiently, the tank
is allowed to drain. However, a considerable quantity of solution, about
400 lbs. to the ton, or, with the 0.25 per cent solution used, about 1 lb.
potassium cyanide to the ton of ore, remains in the vat. To dislodge
this, wash water is used, either plain water or sometimes a weak solution resultant from washing. In the latter case, the weak solution is
stored in separate tanks, and this arrangement allows washing with a
minimum wastage of solution. The extraction of the Mercur ore has
varied; at the beginning of operations it was considerably below 70 per
cent, but as experience with the process increased the results became
better; the average is now given as between 85 and 90 per cent. The cost
of treatment during an early period of the work is given at $2 40 per
ton, divided as follows: Potassium cyanide (1.27 lbs. per ton), 66 cts.;
zinc (0.55 lb. per ton), 5 cts.; labor (seven shifts per twenty-four hours,
6 day and 1 night), $1 12; supplies, repairs, fuel, and freight, 57 cts.;
total (not including office expenses, royalty, and superintendence),
$2 40. Since that period, the expenses have been reduced as the amount
of cyanide lost per ton of ore has been diminished, and a larger quan-

tity of ore is reduced with the same amount of labor. Comparative results by actual experience on the ore by amalgamation and cyanide treatment are as follows: 1,500 tons of ore treated by amalgamation gave an average extraction of 20 per cent at a cost of $4 25 per ton for milling; 1,600 tons of ore, treated during 90 days by cyanide, averaged an extraction of 88.5 per cent, at a cost of $2 25 for milling." The introduction of the cyanide process has made the Mercur Company a remarkable financial success; it paid during the first five months of this year $150,000 in dividends. The plant is being increased to a capacity of 250 tons per day.

(b) **Montana.**—One of the first cyanide mills in this State was erected by F. B. & R. B. Turner, in Revenue, Madison County, who supplied the information which follows: The gold in the Revenue ores has always been very hard to save, the best amalgamation methods only saving from 25 to 27 per cent of the assay-value; the application of the cyanide process increases the returns to from 80 to 87 per cent. Extensive tests have proved that the wet-crushing and cyanide treatment is profitable only on low-grade ore, as the loss of low-grade slimes connected with that method is considered immaterial. The most successful treatment with cyanide has been found to be percolation of dry-crushed ore. The present plant (see diagram, p. 86) is used for wet-crushing of low-grade ore. The tailings from the battery pass into settling pits, and the slimes are allowed to flow into a large reservoir below the mill for eventual future treatment. The tailings settled in the pits are shoveled into the leaching tanks; the running of the pulp direct into the percolation vats proved here, like in many other localities, a failure, owing to the slimes. A dry-crushing plant of 25 tons per day is now in course of construction (see diagram); after its completion, ores of about $25 value will be crushed in the dry, poorer ores in the wet mill. The Revenue ore is nearly pure silica, containing from 1 to 2 per cent of iron peroxide and no sulphurets. It is crushed and passed through a 30-mesh steel screen. Some of the percolation vats are of 10 ft. diameter by a depth of $4\frac{1}{2}$ ft., others are of 12 ft. diameter by 4 ft. deep; they are made of 3-in. Oregon pine. The strength of the solution varies from 0.6 to 1 per cent; percolation takes from 24 to 36 hours, after which the solution, which is circulated or pumped back, varies from 0.4 to 0.8 per cent of cyanide. One half ton of solution, on an average, is used for treating one ton of ore. The total extraction amounts to from 80 to 87 per cent of the assay-value of the ore, 27 per cent of which is obtained from amalgamation on the plates. Cyanide extracts from 73 to 79 per cent of the value left in the tailings after amalgamation. No difficulty is being experienced in precipitating the gold. The consumption of zinc amounts to about half a pound per ton of ore. Sulphuric acid is used for bullion refining; the fineness of melted bullion is from 920 to 940. The total cost of treatment amounts to $5 per ton, including $1 patent-royalty, and consists of the following items: Crushing and labor, $2; chemicals, $2; patent-royalty, $1. The actual consumption of cyanide is from $2\frac{1}{2}$ to 3 lbs. per ton of ore. The total cost of the plant amounted to about $20,000, including engines, boilers, stamps, and vats, all placed in position. Eight to ten men are employed in the mill.

A cyanide mill erected by the Henderson Mountain Mining and

Milling Company, near Cooke, Park County, was worked on surface hematite ore for some time with fairly satisfactory results.

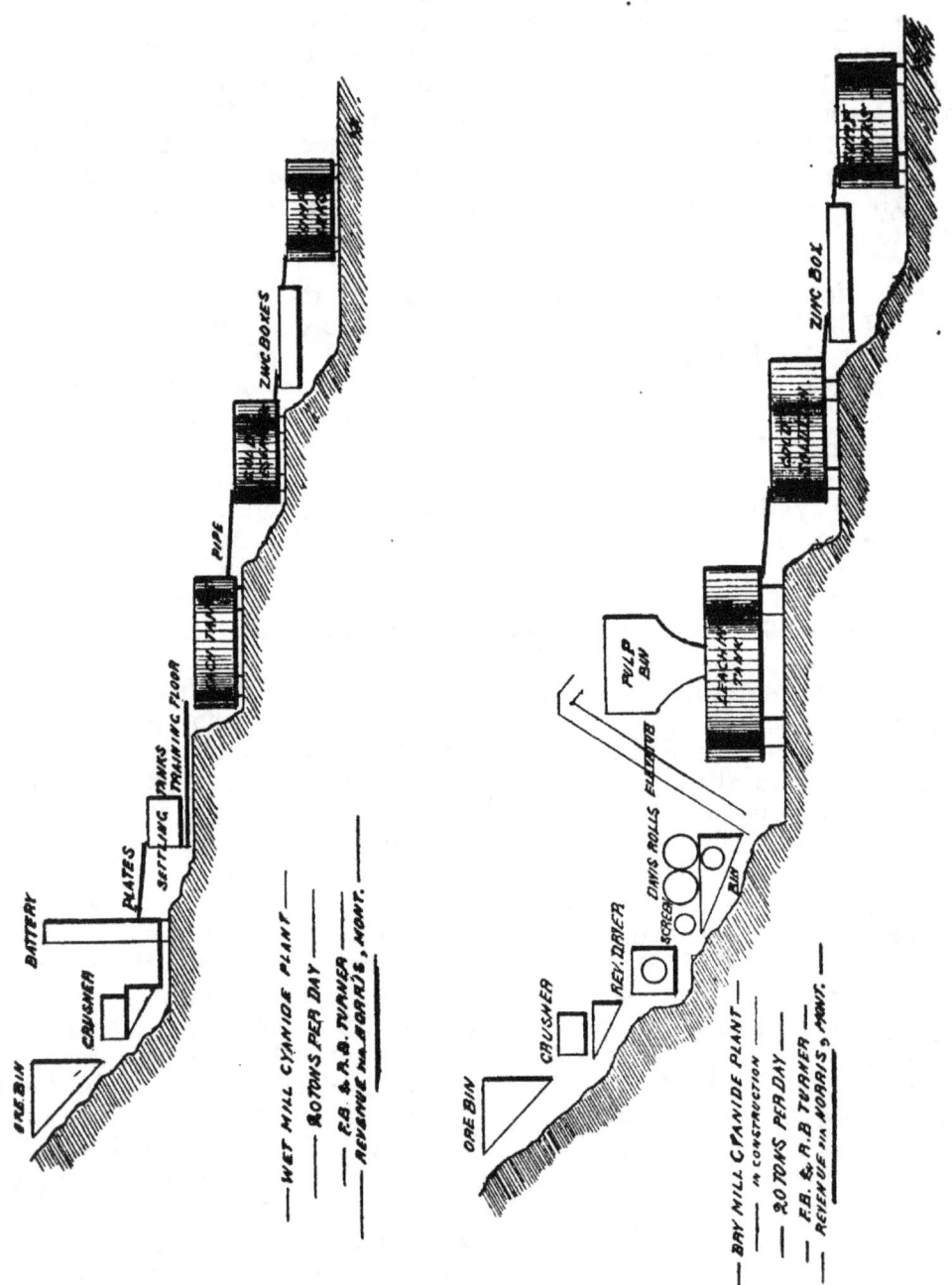

(c) **Colorado.**—The owners of the MacArthur-Forrest patents possess a small plant for testing ore parcels in South Denver.

The Cripple Creek Gold Extraction and Power Company erected a plant in Cripple Creek, in reference to which I received the following communication from the technical manager (Mr. J. K. Turner):

"A small plant has been a remarkable success, and an addition to it is now in course of erection, bringing its capacity up to 50 tons daily. The machinery will finally consist of three Gates' crushers, two pulverizers, four screens, four iron leaching vats of 20 ft. diameter, two solution tanks of 15 ft. diameter, and four zinc boxes of 40 ft. length. The plant is running as a custom mill, and many classes of ore are being treated; at present about 30 tons per day. The ore is crushed dry and passed through a 20-mesh screen. The cyanide solution used is 0.75 per cent, the strength of which is by percolation reduced to 0.5 per cent. One half ton of solution is required for the treatment of one ton of ore. The extraction of gold averages 90 per cent, that of silver 84 per cent. Sulphuric acid is used for bullion refining; the bullion is from 775 to 790 fine. Copper and zinc compounds in •some ores have been found to interfere with extraction. The average value of the ore is $30 before and $3 after cyanide treatment. The cost of treatment amounts to $4 70 per ton. The cost of the plant was $20,000. Seven men are employed in the works."

The successful cyanide treatment of Cripple Creek ores, as here described, is very interesting, for the ores in that district contain a large amount of telluride minerals. Many ores contain sylvanite, krennerite, and calaverite.

The Puzzler Gold Mining and Milling Company of Denver have ceased to work their cyanide plant at Ward, Boulder County, although they were successful with the process.

A cyanide plant at Junction Creek is reported in successful operation. (M. S. P.)

(d) **Nevada.**—A branch company was formed by the owners of the MacArthur-Forrest patents for introducing the cyanide process on the Comstock Lode. A number of laboratory experiments made in the Con. Virginia and California Company did not lead to the adoption of the process.

(e) **Arizona.**—No information can be obtained in reference to the process in this Territory, where a company has been organized for its introduction, with the exception of that obtained from the Champies Mine, Yavapai County, where the results were unsatisfactory, apparently on account of faulty technical manipulation.

(f) **New Mexico.**—The Deep Down Mine had adopted the cyanide process, but abandoned it in favor of pan-amalgamation, the reasons for which are explained (Engineering and Mining Journal) as consisting in a change of the character of the ore.

(g) **South Dakota.**—The Black Hills Gold and Silver Extraction Mining and Milling Company of Deadwood have of late erected a cyanide plant, about which the general manager, Mr. I. S. Childs, gives the following details:

"The ore consists of from 80 to 95 per cent of silica, accompanied by variable quantities of iron, both in the form of peroxide and the various conditions of partial oxidation, and contains traces of copper, manganese, arsenic, and antimony. The ore is dry-crushed with rolls, and passes through a 30-mesh screen. The percolation vats of steel are 24 ft. in diameter by 3 ft. in depth. The cyanide solution is usually of 0.5 per cent strength; its quantity amounts to half the weight of the ore. From 85 to 90 per cent of the assay-value in gold and from 50 to 75 per cent of the silver value is extracted. All extracted bullion is recovered. Lime is used with the ore and caustic soda with the concentrates to remedy the 'acidity.' The consumption of zinc amounts to 0.55 lb. per ounce of bullion recovered. The treatment takes from twenty-four to forty-eight hours. The value of the material before treatment is $20; the tailings assay from $2 to $3 in gold. The total expenses for treatment, including $1 patent-royalty, are $3 50 per ton. The plant for the treatment of 40 tons per twenty-four hours cost $25,000. Fifteen men are employed per twenty-four hours."

The Golden Reward Chlorination Works, of Deadwood, are reported as adding cyanide works to their plant. The works are custom works.

Many tests with cyanide have been made in various other States, in some instances leading to the adoption of the process on a commercial scale; in others, experiments were conducted with little knowledge and in a cursory manner; in other instances, again, the character and qualities of the ore prevented cyanide treatment from being a success. One instance, where the process was first used but afterwards abandoned, is the Creighton Mining and Milling Company, Cherokee County, Georgia, where the extraction from concentrates from old oxidized tailings, as described by the general manager, was reasonably good, being 82 per cent of the assay-value; from the fresh concentrates from deeper levels, however, the returns were only 50 per cent, in consequence of which the process was discarded and barrel chlorination introduced in its place (H. T. Fisher). In other instances the value of the ore in the mine fell off, so that not only the cyanide process, but all operations had to be stopped. An instance in point offers—the Moratock Mine, Montgomery County, North Carolina. The engineer in charge stated (Engineering and Mining Journal) that he had no trouble in treating the ores of that mine by cyanide and making a high extraction (95 per cent) at a moderate cost. Cyanide solution of 0.25 per cent was used; the total cost of mining, milling, and royalty amounted to about $3 75. The mine, however, was shut down, the ore-value receding to $1 25.

(h) California.—The Shasta Gold Recovery Company erected a mill under the direction of Mr. A. B. Paul. The mill has been working for some time successfully. Mr. Paul was one of the first, if not the first, who used wet-crushing of the ore with cyanide solution, instead of water, in the mortars. No authentic information could be obtained in reference to his results.

A cyanide plant has been erected at the Gold Run Mine, Siskiyou County, the working of which has been described in the California Mining Report, No. 11, page 430.

The following information referring to the cyanide process in Kern County, has been contributed by Mr. W. L. Watts, assistant in the field to the California State Mineralogist: "A company which was organized in St. Louis to reopen the Bright Star Mine in Kern County, attempted to work several thousand tons of tailings at that mine by the cyanide process. It is said that the leaching plant used for the purpose could only handle six tons of tailings every twenty-four hours, and that about 45 per cent of the precious metal was recovered."

"*The Cyanide Process at Havilah, Kern County.*—In 1892, Messrs. Stebbins and Porter commenced leaching at Havilah. They first treated

50 tons of heavy sulphuretted concentrates at the Reese mill. The assay-value of these sulphurets ranged from $15 to $48 per ton. Owing to leakage and other causes, their first experiments were not remunerative, but the last and richest portion of this batch of concentrates was treated with great success, 86 per cent of the precious metal being saved. One difficulty encountered arose from the fact that, owing to the sulphurets having long been exposed to the air, they were partially oxidized into acid sulphates; the detrimental effect of these sulphates was eventually overcome by neutralizing them with quicklime. One hundred tons of tailings were then treated at the Hayes mill. These tailings showed an assay-value of $20 per ton, and about 91 per cent of the precious metal was saved. This last lot of tailings consumed 3 lbs. of potassium cyanide to the ton of ore. The process employed is as follows: The tailings are first treated in a tank 6 ft. long, 6 ft. wide, and 6 ft. deep, and holding a charge of six tons of ore. In this tank the tailings are saturated with a solution containing one half of one per cent, by weight, of potassium cyanide. The tailings are allowed to stand in this solution for forty-eight hours, and are then sluiced into 'filter-bottom tanks,' where they are washed. These 'filter-bottom tanks' are 7 ft. long, 10 ft. wide, and 3 ft. deep. The tailings are sluiced from the saturating tank with a stock solution which has a strength of one-fifteenth of one per cent of potassium cyanide. About 1,000 gallons of stock solution are used in sluicing six tons of ore. The filtrate is then tested to determine whether a sufficient percentage of the gold and silver value of the pulp (as shown by assay) has been dissolved. If the filtrate is found to contain a sufficient percentage of the gold and silver, the solution is drawn off and the tailings are washed with fresh water; enough wash-water is used to replace the amount of the solution taken up by the tailings. The amount of water thus added to the solution is usually about 500 gallons to six tons of ore. If on testing the solution it is not found to have dissolved a sufficient percentage of the gold and silver, the stock solution is left on the ore until all the gold and silver which it is possible to extract by this process has passed into solution. The temperature of the solution employed has varied from 40° to 80° Fahr., and within these limits no difference was experienced in the solubility of the gold and silver. In this process it was found to be far better to assay the filtered solution than the macerated pulp, for a perfect volumetric sample of the filtrate can be readily obtained. Test assays—four assay-tons of the auriferous solution of cyanide of potassium are evaporated, and the resultant auriferous compound is mixed with half an assay-ton of litharge, and fluxed with glass. The assay is conducted as an ordinary assay for gold and silver."

"In speaking of the different ores they have treated by the cyanide process, Messrs. Stebbins and Porter state that it has been their experience that the majority of ores showing free gold also contain gold in any sulphide which may be present; but that, except in the case of sulph-arsenides, ores showing no free gold seldom contain auriferous sulphides. In September, 1893, Messrs. Stebbins and Porter were building a mill in which to treat ore by the cyanide process at the Iconoclast Mine."

A cyanide agitation plant was erected by me (the author) in 1893, at the Utica Mine, Calaveras County, which, being successful in all details, shall here be described in full. The Utica Mining Company are the owners of an extensive milling plant, consisting of 160 stamps with Frue vanners and Tulloch concentrators, and a canvas plant for the saving of the sulphuret slimes which escape from the concentrators. Up to last year all concentrates were extracted by the Plattner chlorination process, which treats the coarse vanner-pyrites well; the fine slimes of the canvas plant, however, do not give equally satisfactory results, on account of their containing a very large percentage of carbonate of lime, which is troublesome and involves loss in the roasting furnace on account of its fineness and lightness, and is costly in chlorination on account of its taking up chlorine. The slimes are also difficult to leach in the chlorination vats. I examined the different classes of concentrates as to their fitness for treatment with cyanide. The coarser sulphurets from the concentrators did not give, without further grinding, sufficiently high extraction results to warrant the substitution of the chlorination by the cyanide process. The results of the experiments with the slimes from the canvas plant, which were very satisfactory, led to the construction of the present agitation plant. This plant (see illustration) is completely built of steel and iron, and consists of the following parts: A vertical cylindrical agitator (constructed by Mr. C. D. Lane and myself), 5 ft. in diameter by 5 ft. high, of $\frac{1}{4}$ in. steel plate, with a cast-iron bottom 2 in. thick, with strengthening ribs underneath; to the bottom is cast a cone, through which passes the vertical shaft, which carries four arms. To these are attached the four paddles of $\frac{1}{8}$ in. steel, 6 in. wide, twisted like the plates of a propeller. A ring connecting the four paddle arms gives greater stability to them. The shaft with the paddles can be raised, by means of a screw-spindle, 4 ft. above the bottom of the apparatus. A wrought-iron ring, 3 in. wide and $\frac{1}{2}$ in. thick, riveted outside around the top of the agitator, strengthens the structure. The driving gear is placed below. An opening, 4 in. diameter, in the bottom, discharges the contents of the agitator through a pipe, furnished with a stopcock, onto Scheidel's patent vacuum filter, placed on the floor below and in front of it. Here a perfect separation of the cyanide gold solution is effected from the residues. This filter (see diagram) is built of $\frac{1}{4}$ in. steel, with bottom $\frac{3}{8}$ in. thick; it forms a rectangular box 3 ft. 6 in. deep, 7 ft. long by 5 ft. wide; 2 ft. above the bottom is a perforated steel filter-bottom of $\frac{3}{8}$ in. boiler plate, made in three movable sections, supported by angle-iron running around the sides, and by the vertical support of double T iron. The perforations are of $\frac{1}{2}$ in. diameter, at a distance of $\frac{1}{2}$ in. from each other. This filter-bottom fits closely to the sides of the apparatus; it is covered with a blanket, which is kept in position by bars running along the four sides, and fastened by thumb screws. A grating in three sections of $\frac{3}{8}$ in. round iron serves to protect the cloth; the intervals of 3 in. between the bars are filled in with coarse sand. The filter partition divides the apparatus into two compartments, one above the other; the lower forms a closed box, which is in connection with a duplex vacuum pump, by means of which the air can be rarefied when the filter-bottom is covered with pulp. The upper part, above the filter-bottom, receives the contents of the agitator. The real bottom of the apparatus has a discharge with a 3 in. stopcock, for running off the filtered solution into either one or the other of the two solution tanks, which are standing on the floor one step lower, in front of the filter. All cocks and taps of the plant are of considerable diameter, which secures a quick charge and discharge. The filter is provided with a gauge to indicate the height of the solution within, a gauge indicating the degree of vacuum, an air-tap to permit influx of air when the filtered solution is being discharged, and a manhole.

The mode of working the plant is this: The cyanide solution is charged into the agitator, the paddles are set in motion by revolving the shaft, the ore is charged by degrees, and the agitation is kept up for the required time, after which the pulp is discharged from the agitator onto the filter. The vacuum pump is then set in motion, and filtration under the influence of atmospheric pressure will at once commence. The solution will soon be sucked through; then washing follows, first with liquor from former operations, which has already passed through the zinc boxes, and finally with clear water. These operations of filtering and washing take about two hours. It is advisable to suck the tailings as dry as possible before each new wash is put on, which permits the complete removal of the gold solution with a very small amount of liquid, one half ton of which is sufficient for washing a charge of two tons of ore. If the filtration has been properly managed, no degree of continued washing can improve the results. The filtered solutions are clear. The first or original solution, together with the first wash, will be run off into one of the two solution tanks in front below the vacuum filter; the following washes run into the other. These tanks are 8 ft. long by 3 ft. wide and 3 ft. deep, made of $\frac{1}{4}$ in. steel. Each of the tanks is in connection with a zinc precipitation box, 9 ft. long by 21 in. deep and 9 in. wide, divided into ten compartments; there is an interval of 1 in. between each two compartments. The false perforated steel bottoms of the chambers, which can be removed if desired, are $2\frac{1}{2}$ in. above the true bottom of the box (see diagram, p. 31). The bottom of the box has a number of 1 in. iron faucets, one corresponding with the center of each filter compartment; the sides of the box are 4 in. higher than the partitions within, which insures absolute safety against the liquid running over the sides of the box, if one or the other compartments should become blocked. The gold solution flows into the box through a 1 in. cock, enters the first compartment from below through the perforated false bottom, percolates the zinc shavings placed thereupon, leaves it, and enters the second, and so forth. A steel settling tank, 12 in. deep, 12 in. wide, and 9 ft. 3 in. long, is placed below the precipitating box for receiving the bullion when cleaning up (see general demonstration of process, above). The zinc used consists in turnings of $\frac{1}{8000}$ in. thick, turned from cast zinc cylinders on a lathe; 2 lbs. fill one compartment. The solution passes through the box at the rate of 700 gallons in twenty-four hours. The bullion precipitation of the solution is very efficient as it passes from compartment to compartment, which amounts to passing ten times through a zinc column of 14 in. high by 9 in. square. As shown by the following table of analysis,

One Ton of Liquid contains—	Gold.			Silver.		
	Oz.	Dwts.	Grs.	Oz.	Dwts.	Grs.
Originally	5	14	0	2	4	8
After 14 in. of zinc column		16	1		5	3
After 28 in. of zinc column		5	9		1	10
After 42 in. of zinc column		2	14			17
After 56 in. of zinc column		1	4			12
After 70 in. of zinc column		1	1			11
After 84 in. of zinc column			22			3
After 98 in. of zinc column			15.43			0
After 112 in. of zinc column			14.96			0
After 126 in. of zinc column			13.36			0
After 140 in. of zinc column			12.34			0

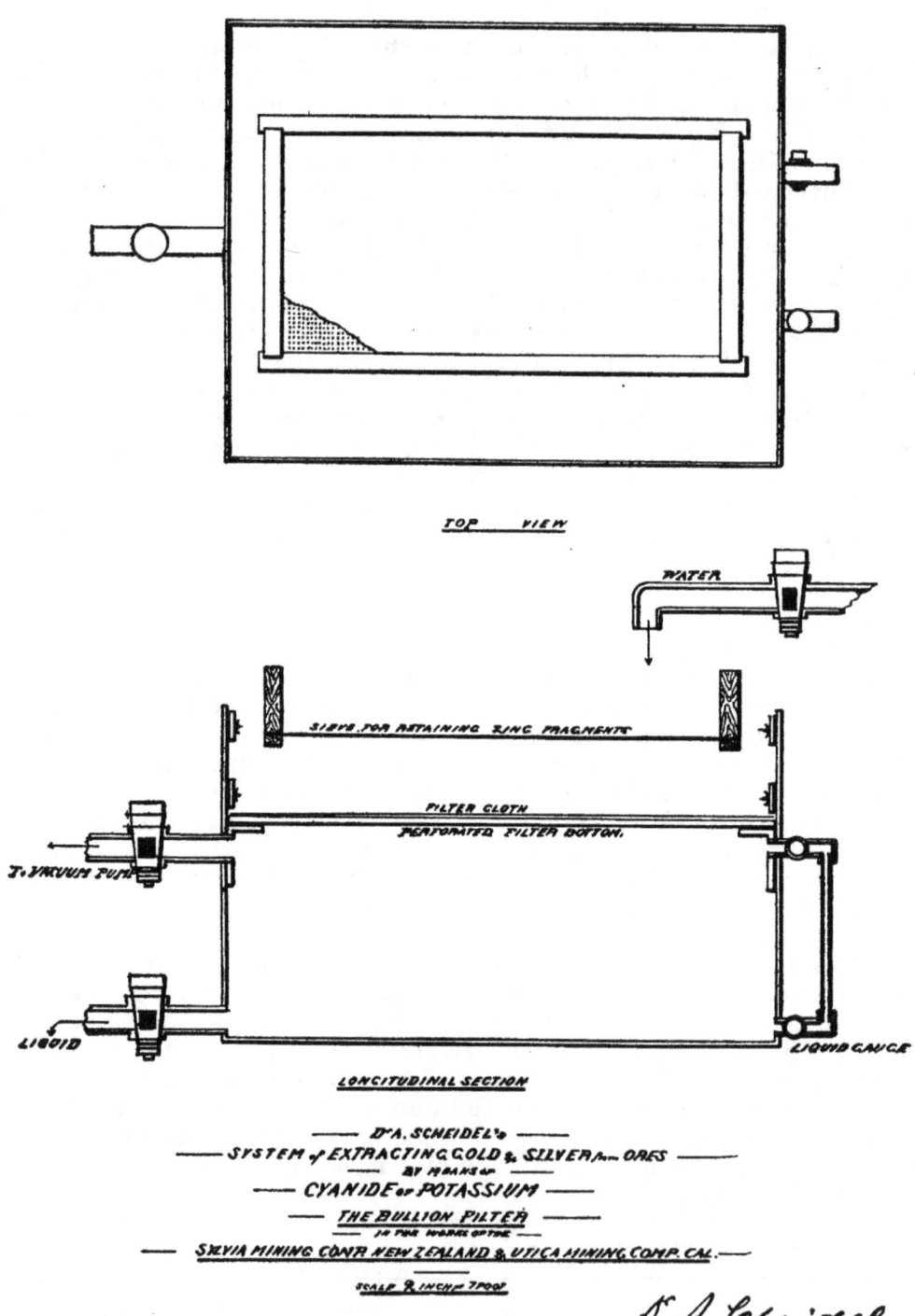

TOP VIEW

WATER

SIEVE FOR RETAINING ZINC FRAGMENTS

FILTER CLOTH

PERFORATED FILTER BOTTOM.

T. VACUUM PUMP

LIQUID

LIQUID GAUGE

LONGITUDINAL SECTION

Dᴿ A. SCHEIDEL's
SYSTEM of EXTRACTING GOLD & SILVER from ORES
BY MEANS of
CYANIDE of POTASSIUM
THE BULLION FILTER
IN THE WORKS OF THE
SILVIA MINING CONᴿ NEW ZEALAND & UTICA MINING COMP. CAL.

SCALE 3 INCH = 1 FOOT

Dᴿ A Scheidel

The solution leaving the zinc box contains only 12.34 grs. of gold, or 0.0045 per cent of its original contents, and only 3 grs. of silver, or 0.0028 per cent of the original silver value. Simultaneously the solutions were analyzed for available cyanide, but no decrease in the strength of the solution, which remained constantly at 0.3185 per cent, could be ascertained. "At another period I studied the solubility of zinc in cyanide solution, of which I give the following figures: 0.2634 grs. of filiform zinc were submerged in 50 cc. cyanide solution of 0.26 per cent; after seven days of frequent agitation these were reduced to 0.2584 grs., and after fifty-six days to 0.2252 grs., which means that after seven days 1.98 per cent, and after fifty-six days 14.47 per cent of the zinc were dissolved. From this observation it follows that the loss of cyanide in the precipitating boxes, by means of its being taken up by zinc, is sometimes overestimated." The washes pass through a similar precipitating box. The liquids, when leaving these boxes, go as liquor No. 1 and liquor No. 2, into tanks of the same size as the solution tanks, from where a pump will deliver them wherever wanted. Liquor No. 1 serves for making up the new solution for the next charge; liquor No. 2 is used for washing purposes on the vacuum filter. No liquor ever leaves the works; the quantity in circulation remains stationary. The bullion obtained from the zinc boxes is passed, as formerly described, through a sieve onto the bullion vacuum, which itself is a miniature reproduction of the vacuum filter (see diagram). It has the following dimensions: Length, 2½ ft.; width, 2 ft.; total depth, 1 ft. 6 in. The perforated filter-bottom is fixed 12 in. above the true bottom. The bullion is very slimy; in fact, it is the more slimy the freer it is from zinc; its filtration and washing take some time. When the mass is tolerably dry, it is put into a wooden tub and treated with diluted sulphuric acid. The heat of the reaction I have always found sufficient to make the operation a speedy and satisfactory one; the bullion is then permitted to settle, the liquid is siphoned off through the bullion filter, and the solid matter is washed by decantation with water. This washing process is continued until all soluble salts are removed. The bullion is then partially dried on the filter, and finally dried in a small muffle furnace; complete drying of the bullion by artificial heat before the acid treatment is not advisable. The thoroughly dry bullion is pulverized and well mixed with soda and borax, and melted in a plumbago crucible as described before. The bullion thus obtained is 946 fine; the slag is clean, it contains the usual few granules of bullion, but, freed from them, does not give any assay results. The bullion could be still further refined, but to no commercial advantage. The steel of the tanks has not as yet shown any effects from cyanide, nor does it exercise any influence on the solutions. All apparatus is composed of plates and sheets riveted together; leakages, if any, can be easily stopped by a varnish made of asphaltum dissolved in bi-sulphide of carbon. As mentioned, this plant has been constructed for the treatment of slime concentrates from the canvas plant; such concentrates contain a varying percentage of carbonate of lime, in some instances as much as 95 per cent, which, however, does not interfere mechanically or otherwise with their satisfactory extraction by cyanide. Such conditions would make chlorination all but impossible, as alluded to before. For agitation the material requires an amount of solution equal to 30 per cent of its weight, and six hours of time. The described plant is capable of treating a much larger amount of slimes than are usually produced per day by the canvas plant; its services are therefore only periodically required. The average consumption of cyanide, calculated from a large tonnage of slimes treated, amounted to 4.3 lbs. per ton, costing $2 27; the labor amounts to $1; the total expenses of treatment by cyanide to $3 50 per ton. The average extraction amounts to 93.18 per cent of the gold, and

90 per cent of the silver assay-value; as high as 96.57 per cent of the gold has been extracted in some instances. The extraction of the gold during the agitation goes on as shown by this table:

Treatment of Slimes by Agitation.	Gold per Ton.	Extraction—Per Cent.
Sample before treatment	$88 00	
Sample after 1 hour's agitation	13 00	85.23
Sample after 2 hours' agitation	11 00	87.50
Sample after 3 hours' agitation	7 00	92.05
Sample after 4 hours' agitation	7 00	92.05
Sample after 5 hours' agitation	6 00	93.18
Sample after 6 hours' agitation	5 00	94.31
Sample after 7 hours' agitation	5 00	94.31
Sample after 8 hours' agitation	5 00	94.31

Within the first hour 85.23 per cent of the gold are extracted; during the following five hours the increase of extraction is slow and irregular; after six hours no further extraction takes place. For experimental purposes I continued agitation up to twelve hours without improving on the result. The treatment of the slime concentrates by agitation was preferred on account of its quicker, cheaper, and better results, as compared with percolation. All Utica concentrates, as in fact all concentrates I ever had to deal with, contain a small amount of amalgam, part of which is found on the bottom of the agitator; part of it leaves the works with the tailings, and is recovered in Hungarian riffles and on amalgamated silver plates; part of the mercury goes undoubtedly into the cyanide solution, and is precipitated with the bullion in the zinc boxes.

Other sulphurets, such as the Frue vanner concentrates of the Utica, Madison, and Eureka mines, were treated on a more or less extensive scale by the same plant; results were, however, not very satisfactory on account of their coarseness. All sulphurets of the Utica Mine are pure sulphide of iron. The fine canvas-plant concentrates, although less clean, are as a rule richer in gold; their extraction averaged 93.18 per cent, whereas the vanner concentrates gave only 81.38 per cent, which, although reasonably good at the rate of $4 per ton cost of treatment, cannot compete with a chlorination treatment, which yields 90 per cent of a $50 ore at a cost of $6 50. (The large size of the Utica chlorination works offers special advantages and permits chlorination at this figure, which is much lower than the cost anywhere else in California.) A large number of tests proved that a high percentage of the gold is contained in the coarser particles of the sulphurets; this will account to some extent for the comparatively low percentage of cyanide extraction. It would lead too far to give here the results of the large number of experiments in reference. The experiments were extended to roasted concentrates after their reduction to uniform size, which gave on a small scale excellent results (98.09 per cent extraction). I am indebted to Colonel Hayward, Mr. Charles D. Lane, and Mr. James Cross, of the Utica Mine, for their permission to publish the diagrams and the described results of the cyanide works which I erected for them. The cost of the plant is divided as follows:

Grading and foundations	$200 00
Building	300 00
Shafting, belting, and putting into place	135 00
Agitator	260 00
Vacuum filter	165 00

Three tanks	$160 00
Two zinc boxes	260 00
Two steel tanks	85 00
One vacuum pump	235 00
One liquid pump	130 00
Pipes, stopcocks, faucets, etc.	70 00
Total	$2,000 00

The Standard Consolidated Mining Company, Bodie, California, started a 100-ton cyanide plant (50-ton vats) on September 17th for the treatment of tailings.

The amount of ores, fine concentrates, and tailings suitable for cyanide treatment is considerable in this State. There is no doubt that a great amount of gold is now being lost with the slimes in most mills here, as elsewhere, and in many instances dry-crushing of the ore and direct cyanide treatment would vastly increase the returns. Coarse gold, if present, can be saved by amalgamation at some later stage of the manipulation.

Generally speaking, the number of commercial successes of the process in the United States is limited, although the number of tests and trials on a smaller or larger scale have been very numerous. The amount of ores suitable for the treatment is large, and the process is becoming more and more an established and acknowledged fact with the mining public. The miners of this country have long been looking for a process for the treatment of low-grade sulphurets which is cheaper in its application than chlorination. Under certain conditions the cyanide process meets the requirements, and will be found particularly valuable in remote places to which the freight expenses are high. The weight of chemicals used in chlorination amounts to about 5 per cent of the ore weight, but only to about 1 per cent in the case of cyanide treatment. By chlorination one ton of chemicals will treat about twenty tons of ore, whereas by cyanide one ton will treat 100 tons; moreover, the cyanide process does not require a special treatment of the ore for the extraction of the silver. No statistics in reference to the gold produced in the United States of America by the cyanide process could be obtained by the author.

D. Mexico, Etc.

A MacArthur-Forrest company is introducing the process in that old mining country. No information, however, could be obtained from that company in reference to their results. In other parts of the world, attempts have been made, chiefly by the owners of the Mac-Arthur-Forrest patents, to introduce the cyanide process. Ores from the republic of Colombia are reported as having been treated with success, and the introduction of the process into that country is now intended. Negotiations for the introduction of the process into the Straits Settlements, Borneo, the mining States of South America, and that great gold-producing country, Russia, are now pending.

I have in this paper been following the cyanide process in its workings and results through the chief places of its application. The process, like most metallurgical processes, has its weak points, which by continued investigations may be strengthened; its strong points are evident from the description of its successful application. Whatever may be the merit of the controversy on the subject of patent-rights in connection with this

process, there is no doubt that Messrs. MacArthur and Forrest deserve
the credit of having first, on a large scale, practically and successfully
applied the cyanide process for working ores.

SUMMARY AND CONCLUSIONS.

Generally speaking, the cyanide process is better suited for the treatment of gold ores than of silver ores, one of the reasons of which may be
found in the great variety of compounds in which the silver occurs, and
a number of these offer difficulties in treatment.   The required long
contact of cyanide, connected with the large consumption of reagent,
prevents the treatment of silver ores in some instances from being a
commercial success, even when chemically a high extraction is obtained.
In reference to the general characteristics of ores which can be successfully treated, it must be said that no definite specification can be given.
The question, Is an ore suited for cyanide treatment? can only be decided
experimentally.   Preliminary but exhaustive experiments on a limited
scale should precede all operations on a large scale.   Many of the problems of the process are of a chemical nature; many of its difficulties are,
however, of a mechanical character, and experience and judgment must
guide in the selection of the plant to make the venture a financial success.   One of the great mechanical difficulties has been, and is still, the
treatment of the slimes, by which is meant the very finest parts and the
clayey portion of ores and tailings.   The final solution of the difficulty
in their treatment will probably be found in dry-crushing the ore and
direct treatment of the crushed material with cyanide.   In cases where
coarse gold is present, the amalgamation of such may be introduced at
some later periods of the manipulation.   Practical experience with the
process extends over only a few years.   It has been found to be well
adapted for free-milling ores with finely divided gold, particularly
so-called float gold, and has given great satisfaction with some pyritic
ores.   Even complex ores containing tellurides have been treated to
advantage.   The process has found its most extensive application on
the Transvaal gold fields; although the average extraction is not
high, it answers there better than any other process attempted for
working the tailings.   In other parts of the gold-producing world, its
application is gaining way by degrees.   Like all metallurgical processes, its success depends on the character of the ore and local circumstances, and failures had to be recorded where these were not
sufficiently considered.   It is certain that our knowledge is as yet
incomplete, and there is still a large amount of ground for the metallurgist and chemist to explore.   We have yet particularly to learn how
to extend the application of the process to more common use.   A comparison of the cyanide process with other processes is a futile task; the
great merit of the treatment is that it comes to fill a want long felt—that
of treating low-grade ores and tailings in a simple and inexpensive way.
It is here that the process antagonizes no other methods, and simply
takes its place in gold metallurgy as a new and powerful means to con-
quer nature.   Wherever the process gives satisfactory results, it offers
great advantages: it does not require roasting furnaces; ores containing
lead, zinc, or earthy carbonates, which cannot be worked to a profit by
chlorination, may be easily and profitably treated by it; as it does not

require smelters, coal, and fluxes, it may be successfully used in remote situations, where smelting is absolutely impossible. One of its great advantages is, it does not require extra treatment for silver, invariably associated with gold in ore. I have generally spoken about gold only in this paper. The remarks referring to it and to its extraction apply with equal force to the silver associated with it. A well-constructed plant and efficient chemical and metallurgical supervision are, however, conditions always necessary to make its application a commercial success. The process is only in its infancy; many of the various and complex problems it has given rise to—such as the reduction in the consumption of cyanide and its regeneration—are still open questions. Its possibilities are great; chemical and mechanical improvements will enlarge the range of its application and usefulness. If it has not proved itself to be the metallurgical panacea that some enthusiasts expected, it has certainly during the four years of its technical application developed into a process of enormous economical importance, and one which justly may be considered a most valuable addition to gold metallurgy.

APPENDIX.

APPENDIX.

UNITED STATES PATENT OFFICE.

JULIO H. REA, OF SYRACUSE, NEW YORK.

IMPROVED MODE OF TREATING AURIFEROUS AND ARGENTIFEROUS ORES.

Specification forming part of Letters Patent No. 61,866, dated February 5, 1867.

To all whom it may concern:

Be it known that I, Julio H. Rae, of Syracuse, in the county of Onondaga and State of New York, have invented a new and useful Improvement in Treating Auriferous and Argentiferous Ores; and I do hereby declare that the following is a full, clear, and exact description thereof, which will enable those skilled in the art to make and use the same, reference being had to the accompanying drawing, forming part of this specification, in which—

Figure 1 represents a transverse vertical central section of the apparatus which may be used in carrying out this invention.

Figure 2 is a plan or top view of the same.

Similar letters of reference in both views indicate corresponding parts.

This invention consists in treating auriferous and argentiferous ores with a current of electricity or galvanism for the purpose of separating the precious metals from the gangue. In connection with the electric current suitable liquids or chemical preparations, such, for instance, as cyanide of potassium, are used, in such a manner that by the combined action of the electricity and of the chemicals, the metal contained in the ore is first reduced to a state of solution and afterwards collected and deposited in a pure state, and that the precious metals can be extracted from the disintegrated rock or ore at a very small expense and with little trouble or loss of time.

In carrying out this process a jar, A, may be used such as shown in the drawing. This jar is made of glass or other suitable material (the size depending upon the electric battery to be used in connection therewith), and into said jar is placed the pulverized rock, filling the same half full or more. On the rock is poured the proper chemical preparation in a fluid state, such, for instance, as cyanide of potassium. Through the center of the jar passes a vertical shaft C, which terminates upon a metal plate, a, by preference of platina, which rests on the bottom of the jar, and to said shaft is attached a cage, B, of platina wire or other suitable material. This cage is made in the form of a truncated cone, its base extending close to the inner circumference of the shaft, or it may be made in any other suitable form or shape, a series of spirals, for instance, which will produce the same effect. On the shaft C is mounted a pulley, E, which may be connected with any suitable mechanism for the purpose of imparting a rotary motion to said shaft and the cage connected therewith, so that the contents of the jar will be agitated and each particle of the pulverized rock shall come in contact with the metal cage B and plate a. The shaft C is connected by a wire, b, with one, say the positive pole of the battery, thus converting the shaft, the cage, and the plate a into an electrode, and the other or negative pole of the battery connects by a wire, c, with a thin slip or coil, d, of copper or other suitable material, forming a base on which the precious metals are deposited. By the action of the electric current the action of the chemicals on the metals contained in the rock is materially facilitated and a perfect solution thereof is effected, and from this solution the precious metals are precipitated upon the base d, whence the same can be easily collected. By this process gold or silver can be extracted from rock in an absolutely pure state and with very little expense.

What I claim as new, and desire to secure by Letters Patent, is—

1. The within-described process of treating auriferous or argentiferous rock by exposing the same to the combined action of a current of electricity and of suitable solvents or chemicals, substantially such as herein specified, or any others which will produce the same effect.

2. Separating gold or silver from the rocks containing the same by the action or aid of electricity, substantially as described.

3. Using the agitator B as an electrode substantially as and for the purpose set forth.

JULIO H. RAE.

Witnesses:
W. HAUFF.
GEO. F. SOUTHERN.

J. H. RAE.

Treating Ores.

No. 61,866.

Patented Feb. 5, 1867.

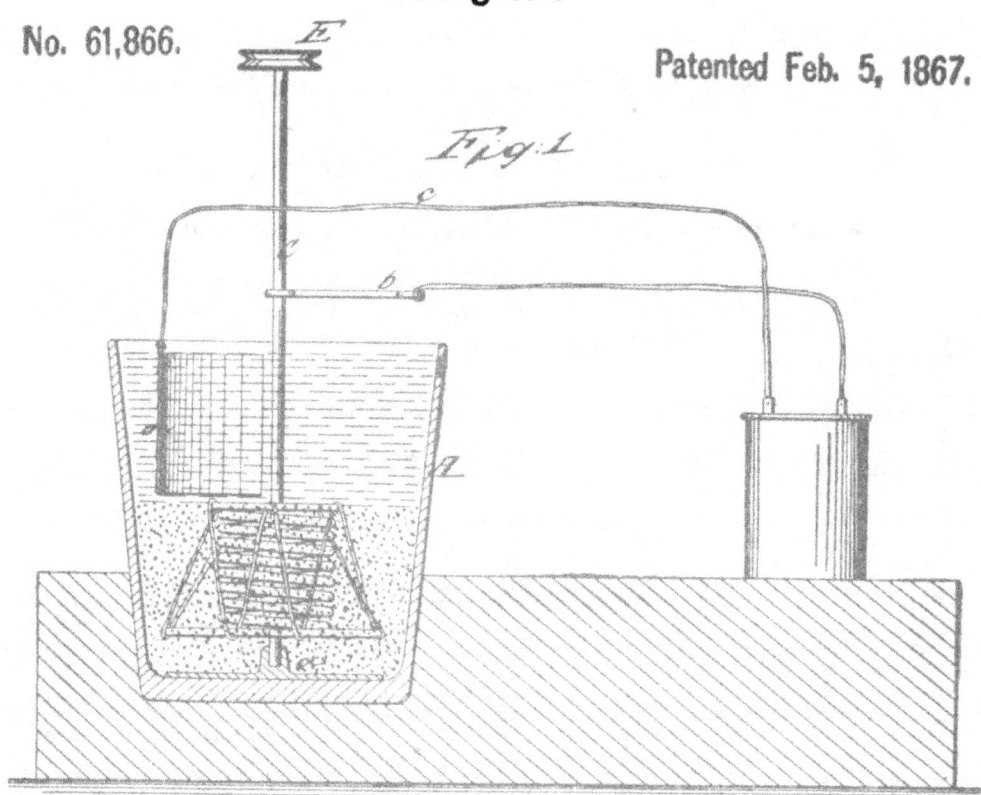

Fig. 1

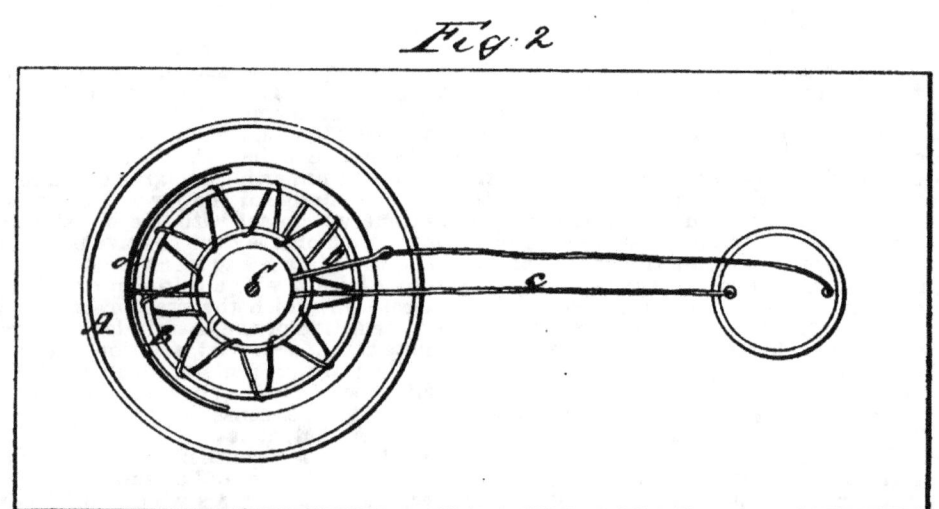

Fig. 2

Witnesses:

Inventor

UNITED STATES PATENT OFFICE.

THOMAS C. CLARK, OF OAKLAND, ASSIGNOR TO JAMES STRATTON, OF SAME PLACE, AND RICHARD E. COADY, OF ALAMEDA, CALIFORNIA, ONE FOURTH TO EACH.

EXTRACTING PRECIOUS METALS FROM ORES.

Specification forming part of Letters Patent No. 229,586, dated July 6, 1880.

(Application filed December 27, 1879.)

To all whom it may concern:

Be it known that I, Thomas C. Clark, of Oakland, county of Alameda, and State of California, have invented an Improvement in Extracting Precious Metals from ores; and I hereby declare the following to be a full, clear, and exact description thereof.

The object of my invention is to perform the disintegration and desulphurization of ores so as to bring the said ore into proper condition for easy pulverization and the precious metals contained therein into a suitable form for amalgamation by freeing them from the union and influence of baser metals. In order to accomplish this object the ore is crushed into pieces about the size of ordinary Indian corn. That portion containing sulphurets generally becomes finer, since it is more friable. The object of crushing it to this size is to prevent loss of gold and to facilitate washing operations.

The ore, after being crushed as described, is placed in an ordinary roasting-furnace. After being roasted for a suitable length of time the heat is raised, so the sulphur will burn freely, after which the heat is let down again, a free supply of oxygen being furnished during the whole process of roasting.

After the ore has become dead and lies like sand in the furnace, and no more scintillation is apparent, it is heated up to a good red heat, but not made too hot.

In a suitable receptacle beside the furnace I form a cold bath, into which the ore is drawn while in its heated condition fresh from the furnace. This bath is formed of a solution of salt, prussiate of potash, and caustic soda or caustic potash.

For one ton of gold ore containing five per cent or less of sulphurets, I form my bath in about the following proportions: I take about thirty gallons of cold water, to which common salt is added until a saturated solution is formed. I then dissolve one pound of prussiate of potash in water and pour it into the solution, and also dissolve one pound of caustic soda in water and add it to the solution. The bath then contains chloride of sodium, prussiate of potash, and caustic soda. For the latter caustic potash may be substituted with a like result.

The red-hot ore being drawn into the cold solution described, a complete desulphurization is effected, as well as a disintegration.

Where there is a higher percentage of sulphur in the ore, additional quantities of the prussiate of potash and caustic soda are added, the proportions of the solution being thus altered to suit the requirements of the class of ore under treatment. The proportions may also be modified for ore of different character.

I am aware that ore has frequently been roasted and dumped while red hot into cold water or into cold solutions, and I therefore do not claim, broadly, such process; but

What I do claim as new, and desire to secure by Letters Patent, is—

The process of disintegrating and desulphurizing ores and freeing the precious metals therein contained, consisting in first roasting the ore to a red heat, and while in that condition placing it in a cold bath composed of a solution of salt, prussiate of potash, and caustic soda or caustic potash, in about the proportions named, substantially as herein described.

In witness whereof I have hereunto set my hand.

THOMAS C. CLARK.

Witnesses:
CHAS. G. YALE.
S. H. NOURSE.

UNITED STATES PATENT OFFICE.

HIRAM W. FAUCETT, of St. Louis, Missouri.

PROCESS OF TREATING ORE.

Specification forming part of Letters Patent No. 236,424, dated January 11, 1881.

(Application filed July 13, 1880. No specimens.)

To all whom it may concern:

Be it known that I, Hiram W. Faucett, a citizen of the United States, residing at St. Louis, in the county of St. Louis and State of Missouri, have invented new and useful Improvements in Process of Treating Ores, of which the following is a specification.

The object of my invention is to treat all refractory ores containing gold and silver for the purpose of separating such metals from the ore.

It is well known that the chemical nature of all refractory ores is more or less of a silicious character, and by desulphurizing or roasting such ores they are rendered porous. Taking advantage of this fact, I subject the silicious ore to a proper chemical bath under pressure, which effectually disintegrates the ore by decomposing or destroying the silica therein, which silica is the chemical agent in the ore which holds or locks the ore together in a compact mass.

To this end my invention consists, broadly, in subjecting hot crushed ores to the action of disintegrating chemicals in solution while under pressure, the pressure being effected by the steam generated by the contact of the hot ore and the chemical solution in a closed vessel.

In carrying out this process I take the ore, crushed as ordinarily for stamp-mill or smelter, and heat the same in any suitable furnace to a sufficient degree and for a proper time to desulphurize it. I then draw the ore, while at red heat, into an iron retort of proper strength to withstand the proposed pressure, and provided with a steam-tight door, and into this retort, through a suitable aperture, after closing the door, I introduce the chemicals in a state of solution, the steam generated creating such pressure within the retort that the chemicals are forced into the silica or rock of the ore, thoroughly disintegrating the same and freeing the metals therefrom, so that the latter are rendered susceptible of ready amalgamation. The process will be facilitated by agitating the retort.

I use different chemicals, according to the different kinds of ore to be treated, and the quantity required depends upon the quantity of silica or other refractory substances to be decomposed to effect a thorough disintegration of the ore.

For the treatment of most of the refractory ores I use chloride of sodium as a base, and in connection therewith nitrate of potash, cyanide of sodium, and about equal parts of sulphate of protoxide of iron and sulphate of copper, the proportions being about as follows, viz.: chloride of sodium, from thirty to forty pounds to the ton; nitrate of potassium, from one to two pounds to the ton; cyanide of sodium, from two to four pounds to the ton; sulphate of protoxide of iron, from one to two pounds to the ton; sulphate of copper, from one to two pounds to the ton. These chemicals are to be dissolved in boiling or hot water of sufficient quantity to cover the ore in the retort. If the ores to be treated are unusually hard or refractory, I add to the above one to two pounds of hydrofluoric acid, or one to two pounds of fluoride of potassium or fluoride of sodium, according to the character of the ore. After the ore has been agitated in the retort a proper time—say from ten to fifteen minutes—under from fifty to one hundred pounds pressure to the square inch, it may be removed while hot to the pulverizer, then passed through any convenient and desired amalgamating process.

It should be here stated that while the ore is in the chemical bath, the latter acts to disintegrate the ore by decomposing or destroying the silica therein, and the ore is thoroughly impregnated with the chemicals, thereby effectually disengaging the particles of metal from the silica, which, if not disengaged, will not amalgamate with the quicksilver in the amalgamating-machines.

To facilitate the carrying out of the process, I prefer to use a cylindrical retort mounted on axial trunnions, in order that it may be rotated for the purpose of agitating its contents. The door of the retort should be in its side, and at each end there should be a projecting coupling-nipple provided with a cut-off cock, to which may be connected pipes, one of which leads from the top, and the other from the bottom, of an elevated steam-tight receiver provided with safety-valve, both pipes being provided with suitable cocks. The chemical solution is then placed in the elevated cylinder, and, after the retort has received its charge of heated ore and been closed, the pipes from the receiver are connected to the coupling-nipples and the cocks all opened. The solution will flow from the bottom of the tank to the retort, and the steam generated in the latter will flow to the top of the cylinder, creating a pressure therein which will force the solution rapidly into the retort. After the pressure has decreased, the cocks may then be closed, the pipes disconnected from the retort, and the latter rotated for the purpose of agitating its contents.

The apparatus thus partially described will form the subject of a separate application for Letters Patent.

I do not confine myself to the chemicals or quantities thereof herein enumerated, as they may be varied as required by the character of the ores to be treated.

I am aware that crushed ores have been subjected to the action of chlorine gas under pressure, and that unroasted pulverized ores have been treated with chemical solutions in a closed vessel under pressure created by the injection of steam, and also that hot roasted ores have been treated by placing cold chemical solutions in contact therewith in the open air, or not under pressure; and I do not claim any of such modes of treatment.

What I claim is—

1. The process herein described for separating metals from ores, the same consisting in subjecting hot crushed ores to the action of disintegrating chemicals in solution under pressure, the pressure being effected by the steam generated by the contact of the hot ores with the chemical solution in a closed vessel, substantially as specified.

2. The process herein described for treating refractory ores for disengaging the precious metals therefrom, the same consisting in subjecting hot crushed ores to the action of a solution of chloride of sodium, nitrate of potash, cyanide of sodium, sulphate of protoxide of iron, and sulphate of copper, under pressure, with or without admixture of hydrofluoric acid, fluoride of potassium, or fluoride of sodium, the pressure being effected by the steam generated by the contact of the solution with the hot ore, substantially as set forth.

In testimony whereof I have hereunto set my hand in the presence of two subscribing witnesses.

HIRAM W. FAUCETT.

Witnesses:
    ROBT. HARBISON,
    JNO. C. ORRICK.

UNITED STATES PATENT OFFICE.

JOHN F. SANDERS, OF OGDEN, UTAH TERRITORY.

COMPOSITION FOR DISSOLVING THE COATING OF GOLD IN ORE.

Specification forming part of Letters Patent No. 244,080, dated July 12, 1881.

(Application filed April 16, 1881. No specimens.)

To all whom it may concern.

Be it known that I, John F. Sanders, of Ogden, in the county of Weber and Territory of Utah, have invented an improved Composition for Dissolving the Coating of Gold in Ore, of which the following is a specification:

The coatings that envelop gold in the ore, and that consist usually of various metallic oxides and of silver, have thus far been difficult to remove, except under the influence of extreme heat, which it is not possible at all places to apply, or by the waste of much valuable time. I have found that a mixture of cyanide of potassium and phosphoric acid, in about the proportions hereinafter mentioned, constitutes a powerful solvent for these coatings of gold ore.

I use in my mixture about sixteen parts of cyanide of potassium to one part glacial phosphoric acid. These two ingredients I mix shortly before the mixture is to be used. The mixture I place into the vessel that contains the covered ore. This vessel preferably is a rotating barrel made of iron or other proper material, and the composition above named is added with sufficient water to form a thick pulp with the raw gravel. The proportions of my improved mixture to the ore will vary, of course, with the varying thickness of covering of the gold. They will, however, be readily ascertained by testing with samples of the ore to be treated. The barrel is rotated or agitated in suitable manner for from fifteen to sixty minutes, as may be required. After agitation the mixture above mentioned will be found, on investigation, to have dissolved the oxides and the sulphurous coatings of the ore, and the agitation of the barrel or vessel removes the dissolved impurities, leaving the gold free and exposed, and permitting it to be amalgamated by the addition of quicksilver, in the usual manner.

The amalgam may be separated from the impurities which have joined with the improved mixture in the manner in which amalgams are usually separated from impurities.

I am aware that cyanides have already been used in the extraction of gold; also, that gold-bearing ores have been disintegrated in the presence of heat by various chemicals. This I do not claim. By using phosphoric acid in the presence of cyanide of potassium I am enabled to dissolve the impurities in a raw state and with great rapidity.

I claim—

The composition of cyanide of potassium and phosphoric acid, in about the proportions mentioned, for the purpose of dissolving the impure coatings of gold, substantially as specified.

JOHN F. SANDERS.

Witnesses:
   WILLY G. E. SCHULTZ,
   WILLIAM H. C. SMITH.

UNITED STATES PATENT OFFICE.

JEROME W. SIMPSON, of NEWARK, NEW JERSEY.

PROCESS OF EXTRACTING GOLD, SILVER, AND COPPER FROM THEIR ORES.

Specification forming part of Letters Patent No. 323,222, dated July 28, 1885.

(Application filed October 20, 1884. No specimens.)

To all whom it may concern :

Be it known that I, Jerome W. Simpson, a citizen of the United States, residing at Newark, in the county of Essex, and State of New Jersey, have invented certain new and useful improvements in processes of extracting gold, silver, and copper from their ores; and I do hereby declare the following to be a full, clear, and exact description of the invention, such as will enable others skilled in the art to which it appertains to make and use the same.

The object of this invention is to extract certain metals from their ores more effectually and at a reduced cost; and it consists in the processes hereinafter set forth, and finally embodied in the clauses of the claims.

To carry my invention into effect, I first grind or crush the ore containing the metal to be extracted to a powder of more or less fineness. This powder is then treated with certain salts in solution adapted to combine chemically with the metal in said ore and form therewith a soluble salt. After thorough agitation to mix the solution with the ore, the mixture is allowed to stand until the solid matter is settled and the solution has become clear. I then suspend a piece or plate of zinc therein, which causes the metal dissolved in the salt solution to be precipitated thereon, from which it can be removed by scraping or by dissolving the zinc in sulphuric or hydrochloric acid. The precipitated metal may then be melted into a button.

The salt solution I use for dissolving the metal from the ore is composed of one pound of cyanide of potassium, one ounce carbonate of ammonia, one half ounce chloride of sodium, and sixteen quarts of water, or other quantities in about the same proportions.

This solution is particularly adapted for ores containing gold, silver, and copper in the form of sulphurets.

For an ore containing gold and copper only I use cyanide of potassium and carbonate of ammonia about in the proportions named.

For ores rich in silver I employ a proportionately larger quantity of chloride of sodium.

I am aware that cyanide of potassium, when used in connection with an electric current, has been used for dissolving metal, and also that zinc has been employed as a precipitant, and the use of these I do not wish to be understood as claiming, broadly.

I am also aware that carbonate of ammonia has been employed for dissolving such metals as are soluble in a solution thereof, and the use of this I do not claim; but

What I claim as new is—

1. The process of separating gold and silver from their ores, which consists in subjecting the ore to the action of a solution of cyanide of potassium and carbonate of ammonia, and subsequently precipitating the dissolved metal, substantially as set forth

2. The process of separating metals from their ores, to wit: subjecting the ore to the action of a solution of cyanide of potassium, carbonate of ammonia, and chloride o sodium, and subsequently precipitating the dissolved metals.

In testimony that I claim the foregoing, I have hereunto set my hand this 15th day of October, 1884.

JEROME W. SIMPSON.

Witnesses:
 OLIVER DRAKE.
 CHARLES H. PELL.

[Fifth Edition.]

No. 14,174.   A. D. 1887.

Date of application, 19th Oct., 1887; complete specification left, 26th July, 1888—Accepted 10th Aug., 1888.

PROVISIONAL SPECIFICATION.

IMPROVEMENTS IN OBTAINING GOLD AND SILVER FROM ORES AND OTHER COMPOUNDS.

We, John Stewart MacArthur, Analytical Chemist, of 15 Princes Street, Pollokshields, in the County of Renfrew, North Britain, Robert Wardrop Forrest, M.D., and William Forrest, M.B., both of 319 Crown Street, Glasgow, in the County of Lanark, North Britain, do hereby declare the nature of this invention to be as follows:

This invention has principally for its object the obtaining of gold from its ores or other compounds, but it is also applicable for obtaining silver from its ores or compounds; and it comprises an improved process, which, whilst applicable to ores or compounds generally, is effectual with ores and compounds from which gold or silver have hitherto not been easily obtainable.

In carrying out the invention the ore or other compound in a powdered state is treated with a solution containing cyanogen or a cyanide (such as the cyanides of potassium, sodium, or ammonium), or other substance or compound containing or yielding cyanogen, till all or nearly all of the gold and the silver are dissolved; the operation being conducted in a wooden vessel or a vessel made of or lined with a material not acted on to any considerable extent by the solution or substances contained therein. The solution is then drawn off and the metal or metals are recovered by any suitable process, and the cyanogen, cyanide, or substance containing or yielding cyanogen may be regenerated. The cyanogen or substance containing or yielding cyanogen may be used as such, or such materials may be taken as will by mutual action form cyanogen or substances containing or yielding same.

Under certain circumstances it may be found desirable to conduct the operation under pressure, in which case a closed vessel must be employed, and in any case, if found advisable, such operation may be carried on under varying conditions of temperature.

Dated this 19th day of October, 1887.

ALLISON BROS.,
Agents for the Applicants.

COMPLETE SPECIFICATION.

Improvements in Obtaining Gold and Silver from Ores and Other Compounds.

We, John Stewart MacArthur, Analytical Chemist, of 15 Princes Street, Pollokshields, in the County of Renfrew, North Britain, Robert Wardrop Forrest, M.D., and William Forrest, M.B., both of 319 Crown Street, Glasgow, in the County of Lanark, North Britain, do hereby declare the nature of this invention, and in what manner the same is to be performed, to be particularly described and ascertained in and by the following statement, that is to say:

This invention has principally for its object the obtaining of gold from its ores or other compounds, but it is also applicable for obtaining silver from its ores or compounds; and it comprises an improved process, which, whilst applicable to ores or compounds generally, is effectual with ores and compounds from which gold or silver have hitherto not been easily obtainable because of the presence of various other metals or their compounds, or because of the physical or chemical condition of the gold or silver in the ores or compounds.

In carrying out the invention the ore or other compound in a powdered state is treated with a solution containing cyanogen or cyanide (such as cyanide of potassium, or of sodium, or of calcium), or other substance or compound containing or yielding cyanogen. In practice we find the best results are obtained with a very dilute solution, or a solution containing or yielding an extremely small quantity of cyanogen or a cyanide, such dilute solution having a selective action such as to dissolve the gold or silver in preference to the baser metals. In preparing the solution we proportion the cyanogen to the quantity of gold or silver or gold and silver estimated by assay or otherwise to be in the ore or compound under treatment, the quantity of a cyanide or cyanogen-yielding substance or compound being reckoned according to its cyanogen. We mix the powdered ore, or compound, with the solution in a vessel made of or lined with wood or any other convenient material not appreciably acted on by the solution. The process is expedited by stirring the mixture of ore and solution intermittently, or continuously for which purpose any convenient mechanical agitator may be fitted to the vessel When all or nearly all the gold or silver is dissolved the solution is drawn off from the ore or undissolved residue, and is treated in any suitable known way, as for example with zinc, for recovering the gold and silver. The residuary cyanogen compounds may also be treated by known means for regeneration or reconversion into a condition in which they can be used for treating fresh charges of ores or compounds.

Any cyanide soluble in water may be used, such as ammonium, barium, calcium, potassium, or sodium cyanide, or a mixture of any two or more of them, or any mixture of materials may be taken which will, by mutual action, form cyanogen, or a substance or substances containing or yielding cyanogen.

In dealing with ores or compounds containing, per ton, twenty ounces or less of gold or silver, or gold and silver, we generally use a quantity of cyanide, the cyanogen of which is equal in weight to from one to four parts in every thousand parts of the ore or compound, and we dissolve the cyanide in a quantity of water of about half the weight of the ore. In the case of richer ores or compounds, whilst increasing the quantity of cyanide to suit the greater quantity of gold or silver, we also increase the quantity of water so as to keep the solution dilute. In using free cyanogen, the cyanogen obtained as a gas in any well known way is led into water to form the solution to be used in our process; or any suitable known mode of setting cyanogen free in solution may be employed.

In some circumstances it may be found desirable to conduct the operation under pressure in a closed vessel; and a higher than the ordinary temperature may be used if found desirable.

Having now particularly described and ascertained the nature of our said invention and in what manner the same is to be performed, we declare that what we claim is—

1. The process of obtaining gold and silver from ores and other compounds, consisting in dissolving them out by treating the powdered ore or compound with a solution containing cyanogen or a cyanide or cyanogen-yielding substance, substantially as hereinbefore described.

2. The process of obtaining gold and silver from ores and other compounds, consisting in dissolving them out by treating the powdered ore or compound with a dilute solution containing a quantity of cyanogen or a cyanide or cyanogen-yielding substance, the cyanogen of which is proportioned to the gold or silver or gold and silver, substantially as hereinbefore described.

Dated this 16th day of July, 1888.

ALLISON BROS.,
Agents for the Applicants.

UNITED STATES PATENT OFFICE.

———

JOHN STEWART MacARTHUR, of Pollokshields, County of Renfrew, and ROB-ERT W. FORREST and WILLIAM FORREST, of Glasgow, County of Lanark Scotland.

PROCESS OF OBTAINING GOLD AND SILVER FROM ORES.

———

Specification forming part of Letters Patent No. 403,202, dated May 14, 1889.

(Application filed November 9, 1887. Serial No. 254,699. (No specimens.) Patented in England, October 19, 1887, No. 14,174; in Cape of Good Hope, January 7, 1888, No. 6–101; in Victoria, January 19, 1888, No. 5,572; in New South Wales, January 21, 1888, No. 453; in South Australia, January 23, 1888, No. 948; in Tasmania, January 24, 1888, No. 529; in New Zealand, February 1, 1888, No. 2,775; in Canada, February 6, 1888, No. 28,471; in France, April 6, 1888, No. 189,808; in Belgium, July 24, 1888, No. 82,673; in Brazil, August 23, 1888, No. 619; in Portugal, August 30, 1888, No. 1,272; in Italy, September 30, 1888, No. 23,852, and in Spain, October 2, 1888, No. 8,538.)

To all whom it may concern:

Be it known that we, John Stewart MacArthur, a subject of the Queen of Great Britain, residing at 15 Princes Street, Pollokshields, in the County of Renfrew, Scotland, and Robert Wardrop Forrest and William Forrest, both subjects of the Queen of Great Britain, residing at 319 Crown Street, Glasgow, in the County of Lanark, Scotland, have invented certain new and useful Improvements in Processes of Obtaining Gold and Silver from Ores (for which we have obtained patents in the following countries: Great Britain, No. 14,174, dated October 19, 1887; Cape of Good Hope, No. 6–101, dated January 7, 1888; Victoria, No. 5,572, dated January 19, 1888; New South Wales, No. 453, dated January 21, 1888; South Australia, No. 948, dated January 23, 1888; Tasmania, No. 529, dated January 24, 1888; New Zealand, No. 2,775, dated February 1, 1888; Canada, No. 28,471, dated February 6, 1888; France, No. 189,808, dated April 6, 1888; Belgium, No. 82,673, dated July 24, 1888; Brazil, No. 619, dated August 23, 1888; Portugal, No. 1,272, dated August 30, 1888; Italy, No. 23,852, dated September 30, 1888, and Spain, No. 8,538, dated October 2, 1888); and we do hereby declare that the following is a full, clear, and exact description of the invention, which will enable others skilled in the art to which it appertains to make and use the same.

This invention has principally for its object the obtaining of gold from ores; but it is also applicable for obtaining silver from ores containing it, whether with or without gold, and it comprises an improved process which, while applicable to auriferous and argentiferous ores generally, is advantageously and economically effective with refractory ores, or ores from which gold and silver have not been satisfactorily or profitably obtained by the amalgamating or other processes hitherto employed, such as ores containing sulphides, arsenides, tellurides, and compounds of base metals generally, and ores from which the gold has not been easily or completely separable on account of its existing in the ores in a state of extremely fine division.

The invention consists in subjecting the auriferous or argentiferous ores to the action of a solution containing a small quantity of a cyanide, as hereinafter set forth, without any other chemically-active agent, such quantity of cyanide being reckoned according to its cyanogen, and the cyanogen being proportioned to the quantity of gold or silver, or gold and silver, estimated by assay or otherwise to be in the ores under treatment. By treating the ores with the dilute and simple solution of a cyanide the gold or silver is, or the gold and silver are, obtained in solution, while any base metals in the ores are left undissolved, except to a practically inappreciable extent, whereas when a cyanide is used in combination with an electric current or in conjunction with another chemically-active agent—such as carbonate of ammonium, or chloride of sodium, or phosphoric acid—or when the solution contains too much cyanide, not only is there a greater expenditure of chemicals in the first instance, but the base metals are dissolved to a large extent along with the gold or silver, and for their subsequent separation involve extra expense, which is saved by our process.

In practically carrying out our invention we take the ore in a powdered state and mix with it the solution of a cyanide in a vessel made of or lined with any material not appreciably acted on by the solution. We employ a vessel made of or lined with wood; but it may be made of or lined with any ordinary inert material—such as stone, brick, slate, rubber, gutta-percha, cement, glass, earthenware, iron (plain, tinned, or enameled), or lead. The process is expedited by stirring or triturating the mixture of ore and solution intermittently or continuously, for which purpose any convenient mechanical agitator may be fitted to the vessel. A pan-mill with edge runners or other known triturating device may be advantageously used. The solution is allowed to act on the ore until the gold or silver is all or nearly all dissolved, and the solution is then drawn off from the ore or undissolved residue.

Any cyanide soluble in water may be used—such as ammonium, barium, calcium, potassium, or sodium cyanide, or a mixture of any two or more of them. We regulate the quantity of cyanide so that its cyanogen will be in proportion to the quantity of gold or silver or gold and silver in the charge of ore; but in all cases we dissolve it in sufficient water to keep the solution extremely dilute, because it is when the solution is dilute that it has a selective action such as to dissolve the gold or silver in preference to the baser metals.

In dealing with ores containing per ton twenty ounces or less of gold or silver or gold and silver, we find it most advantageous to use a quantity of cyanide the cyanogen of which is equal in weight to from one to four parts for every thousand parts of the ore, and we dissolve the cyanide in a quantity of water of about half the weight of the ore. We generally use a solution containing two parts of cyanogen for every thousand parts of the ore. In the case of richer ores, while increasing the quantity of cyanide to suit the greater quantity of gold or silver, we also increase the quantity of water so as to keep the solution dilute. In other words, the cyanide solution should contain from two to eight parts, by weight, of cyanogen to one thousand parts of water, and the quantity of the solution used should be determined by the richness of the ore. After the solution has been decanted or separated from the undissolved residues the gold and silver may be obtained from it in any convenient known way—such as evaporating the solution to dryness and fusing the resulting saline residue, or by treating the solution with sodium amalgam.

Having fully described our invention, what we desire to claim and secure by Letters Patent is—

The process of separating precious metal from ore containing base metal, which process consists in subjecting the powdered ore to the action of a cyanide solution containing cyanogen in the proportion not exceeding eight parts of cyanogen to one thousand parts of water.

JOHN STEWART MacARTHUR.
ROBT. W. FORREST.
W. FORREST.

Witnesses:
ROBT. DUNLOP,
WILLIAM BRUNTON,
Law clerks, both of 160 West George Street, Glasgow.

UNITED STATES PATENT OFFICE.

JOHN STEWART MacARTHUR, OF POLLOKSHIELDS, COUNTY OF RENFREW, SCOTLAND.

METALLURGICAL FILTER.

Specification forming part of Letters Patent No. 418,138, dated December 24, 1889.

(Application filed November 13, 1889. Serial No. 330,195. No model.)

To all whom it may concern:

Be it known that I, John Stewart MacArthur, a subject of the Queen of Great Britain, residing at Pollokshields, in the county of Renfrew, Scotland, have invented a new and useful Improvement in Metallurgical Filters, of which the following is a specification:

This invention relates to a filter for precipitating and separating precious metals from solutions containing them—such, for instance, as chlorides, bromides, theosulphates (sometimes called "hyposulphites"), or sulphates obtained in the well-known Plattner, von Patera, Russell, Ziervogel, and Augustine extracting processes.

The object of the invention is to provide a filter having a large active surface for the metals in solution.

In the accompanying drawings, Figure 1 is a sectional elevation of a series of these improved filters. Fig. 2 is a longitudinal vertical section of a filtering apparatus comprising two of these improved filters constructed in modified form. Fig. 3 represents a zinc filiform sponge, constituting the principal feature of this improved filter, the filaments of the sponge being represented on an enlarged scale.

Similar numerals of reference indicate corresponding parts in the different figures.

This improved filter comprises a containing-vessel 10 and a zinc sponge 11, disposed therein. The zinc sponge is preferably supported on a perforated false bottom 12, disposed within said vessel near the bottom proper thereof. The vessel is provided with an inlet-tube 13 and an outlet-tube 14, the inlet-tube being preferably disposed near the bottom of the vessel and the outlet-tube near the top thereof, each of said tubes being provided with a coupling-nut 15 when the vessels are used in series.

A number of these filters are preferably arranged in series, as represented in Fig. 1, from six to ten being ordinarily employed. When so arranged, the filters are connected by pipes 16, which extend from the outlet near the top of one vessel to the inlet near the bottom of the adjacent vessel. A reservoir or tank 17 for containing the solution holding the precious metals is disposed adjacent to the first filter of the series and elevated a sufficient distance to secure a proper flow of the liquid through the filters. This tank is provided with an outlet-tube 18 near its bottom, said tube being provided with a stop-cock 19, and connected by pipe 20 with the inlet-tube of the first filter of the series. The zinc sponge, which constitutes the filtering material and precipitant, is preferably composed of fine threads or filaments of zinc interlocked together. The zinc threads from which the sponge is formed are cut by a turning tool from a series of zinc disks held between lathe-centers and turned; or the zinc sponge may be formed by passing molten zinc, at a temperature just above the melting-point, through a fine sieve and allowing it to fall into water. This improved zinc sponge presents a very large contact-surface for the action of the solution, and it does not become easily choked. Each containing-vessel may be provided with a vertical partition or partitions 21, as illustrated in Fig. 2, whereby the vessel is divided into two or more compartments or filtering-chambers. These partitions extend to a point near the bottom or top of the vessel, as the case may be, or they are provided with holes near the top or near the bottom of the vessel. In case the vessel has three or more filtering-chambers, the partitions are provided with communicating-openings, disposed alternately near the bottom and top of the vessel, whereby the passage of the solution is downward through one of the filtering-chambers, upward through the adjoining filtering-chamber, and downward again through the third filtering-chamber, and so on.

In the use of this improved filtering apparatus the solution containing the precious metal is placed in the tank 17 and the cock 19 is opened. In case a series of separate filters is employed, as represented in Fig. 1, the solution passes from the tank through the pipe 20 and into the first filter of the series, near the bottom thereof, beneath the false bottom 12, thence upward through the zinc sponge within the filter, thence outward near the top of the first filter, thence through the connecting-pipe to the next filter of the series, where it again enters near the bottom and passes upward through the zinc sponge to the top of the second filter of the series, and so on. The metal which is not precipitated by the first filter is caught in the zinc sponge of the succeeding filters of the series.

In case filters having a number of compartments are employed, the solution is preferably admitted to the first compartment at the top thereof, and passes down through the zinc sponge contained in said compartment to near the bottom thereof, and thence passes into the second compartment and upward through the zinc sponge therein contained to near the top of said compartment, and thence downward through the next

(No Model.)

J. S. MacARTHUR.
METALLURGICAL FILTER.

No. 418,138.   Patented Dec. 24, 1889.

Fig 1.

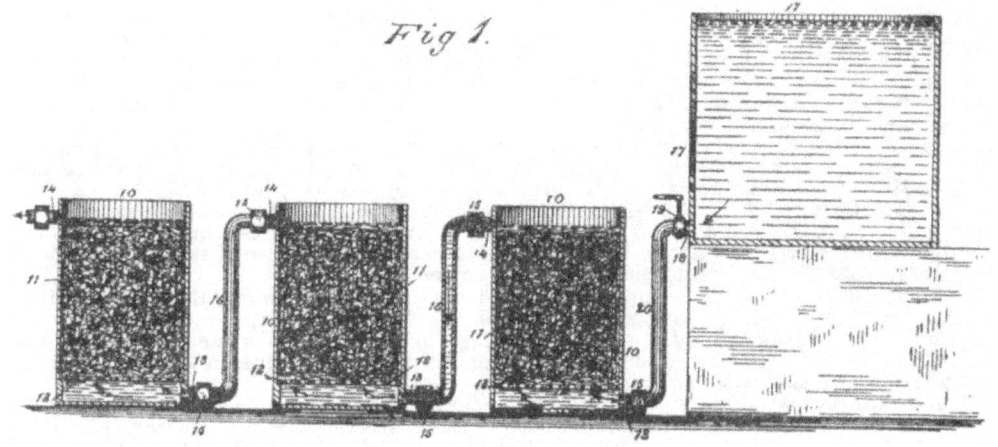

Fig 2.

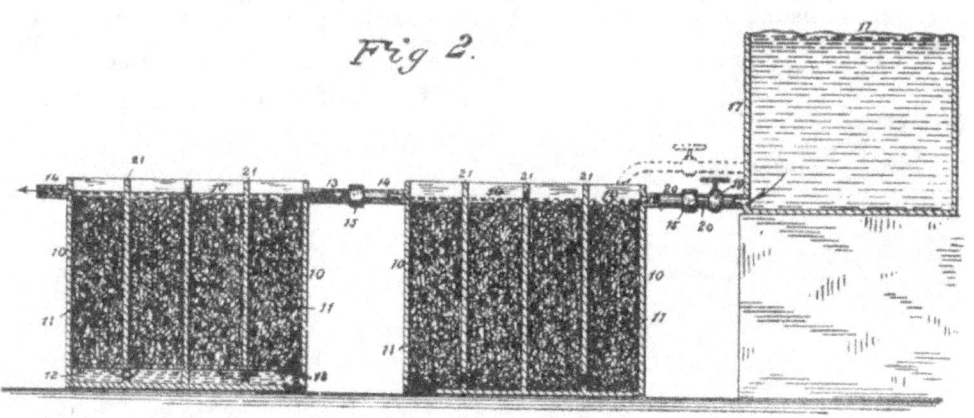

Fig 3.

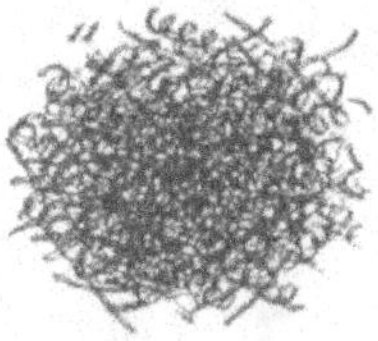

WITNESSES .
Harry King
C. A. Weed.

INVENTOR
John Stewart MacArthur
By F. C. Somes,
Attorney

8CP

compartment, and so on through the several compartments of the compound filter, and thence to the next compound filter of the series and through its several compartments. The precious metal may be separated from the zinc sponge by distillation, or the zinc sponge containing the precious metal may be placed in a suitable sieve and subjected to a screening operation, preferably under water. In this operation the greater part of the precious metal will pass through the sieve and the greater part of zinc sponge will remain therein.

I claim as my invention—

1. A metallurgical filter for separating a precious metal from a solution containing said metal, consisting of a vessel provided with inlet and outlet openings, and a zinc sponge disposed in said vessel between said openings, substantially as described.

2. A metallurgical filter for separating a precious metal from a solution containing said metal, consisting of a vessel provided with inlet and outlet openings and a filiform zinc sponge disposed in said vessel between said openings, substantially as described.

3. A metallurgical filter for separating a precious metal from a solution containing said metal, consisting of a vessel provided with a perforated false bottom, a zinc sponge within said vessel above said false bottom, and inlet and outlet openings above and below said sponge, substantially as described.

4. A metallurgical filter for separating a precious metal from a solution containing said metal, consisting of a vessel provided with a perforated false bottom, a filiform zinc sponge within said vessel above said false bottom, and inlet and outlet openings above and below said filiform sponge, substantially as described.

5. A metallurgical filtering apparatus for separating a precious metal from a solution containing said metal, consisting of a series of vessels, a zinc sponge in each of said vessels, pipes connecting the outlet-tube of one vessel of the series with the inlet-tube of the adjacent vessel of the series, and a reservoir for supplying the solution to the first vessel of the series, substantially as described.

6. A metallurgical filtering apparatus for separating a precious metal from a solution containing said metal, consisting of a series of vessels, each of which has an inlet-tube near its bottom, an outlet-tube near its top, and a perforated false bottom above the inlet-tube, zinc sponges disposed in the several vessels, pipes connecting the inlet and outlet-tubes of the several vessels, and a reservoir for supplying the solution to the first vessel of the series, substantially as described.

7. A metallurgical filter for separating a precious metal from a solution containing said metal, consisting of a vessel provided with a partition dividing said vessel into a plurality of filtering-chambers, said partition being provided with openings near one end, and zinc sponges disposed in said compartments, substantially as described.

JOHN STEWART MacARTHUR.

Witnesses:
  F. C. Somes.
  Gordon Wilson, Jr.

UNITED STATES PATENT OFFICE.

JOHN STEWART MacARTHUR, of Pollokshields, County of Renfrew, and ROBERT WARDROP FORREST and WILLIAM FORREST, of Glasgow, County of Lanark, Assignors to the CASSEL GOLD EXTRACTING COMPANY (Limited) of Glasgow, Scotland.

PROCESS OF SEPARATING GOLD AND SILVER FROM ORE.

Specification forming part of Letters Patent No. 418,137, dated December 24, 1889.

(Application filed April 4, 1889. Serial No. 305,998. [Specimens.] Patented in Natal, September 11, 1888, No. 32; in New South Wales, September 27, 1888, No. 965, and in Tasmania, September 29, 1888, No. 609.)

To all whom it may concern:

Be it known that we, John Stewart MacArthur, residing at Pollokshields, in the county of Renfrew, and Robert Wardrop Forrest and William Forrest, both residing at Glasgow, in the county of Lanark, Scotland, all subjects of the Queen of Great Britain, have invented certain new and useful improvements in the process of separating gold and silver from ores (for which we have received Letters Patent in Natal, No. 32, dated September 11, 1888; New South Wales, No. 965, dated September 27, 1888, and Tasmania, No. 609, dated September 29, 1888); and we do hereby declare that the following is a full, clear, and exact description of the invention, which will enable others skilled in the art to which it appertains to make and use the same.

This invention relates to an improvement in the process of separating precious metals from ores described in Letters Patent of the United States, No. 403,202, granted to us May 14, 1889. In that process a cyanide is used as the separating agent, and it has been found that ores containing pyrites or sulphurets which have been exposed to the weather and become partially oxidized absorb a comparatively large quantity of the cyanide.

The object of this invention is to economize the process by preventing the absorption of the cyanide.

The invention consists in separating precious metals from ores by first neutralizing the ore by the addition of an alkali or alkaline earth and then leaching such prepared charge with a cyanide solution.

In carrying out the first or preparatory step of this improved process, we take ore containing iron pyrites or other compound which has become partially oxidized by exposure to the weather and mix with it, when in a powdered state, a sufficient quantity of potash, lime, or other alkali or alkaline earth, to neutralize the salts of iron or other objectionable ingredients formed by the partial oxidation.

The quantity of alkali or alkaline earth to be employed will depend upon the nature of the ore, and must be determined by first taking a test quantity of the particular ore to be treated and adding the alkali or alkaline earth thereto until the alkali ceases to be absorbed. When this condition is reached the liquid will cause red litmus-paper to turn blue. The proportion of the alkali or alkaline earth so absorbed will indicate the proper proportion thereof to be added to the bulk of the ore to be treated. In case lime is employed, 1 per cent of the alkali to 99 per cent of ore will generally be found sufficient. After this preparatory treatment the ore (which may consist of tailings or residues from other processes or operations) is treated with the cyanide solution by being agitated therewith or by being ground therewith in a pan-mill or other suitable grinding-mill; or, as we find preferable in the case of some ores, the cyanide solution may be made to percolate through said ores one or more times until all or nearly all the precious metals are dissolved. For this percolation very simple tanks, vats, or vessels may be used, such vessels being provided with permeable false bottoms or any suitable filtering apparatus. The cyanide solution containing the gold or silver is next made to pass through a sponge of zinc, whereby said metal is precipitated from the solution and retained in the sponge. The zinc sponge is preferably composed of fine threads or filaments of zinc. These zinc threads are formed in shavings cut by a turning tool from a series of zinc disks held in a lathe; or the sponge may be formed by passing molten zinc at a temperature just above the melting point through a fine sieve and allowing it to fall into the water. The sponge thus formed presents a very large contact surface for the solution, and it does not become easily choked.

The precious metals may be separated from the zinc sponge by distillation; or the zinc sponge containing the precious metal may be placed in a suitable sieve and subjected to a screening operation, preferably under water. In this operation the greater part of the precious metal will pass through the sieve and the greater part of the zinc sponge will remain therein.

We claim as our invention—

1. The process of separating precious metals from an ore, which consists in neutralizing the ore by the addition of an alkali or alkaline earth, and then leaching the neutralized ore with a cyanide solution.

2. The process of separating precious metal from an ore, which consists in neutralizing the ore by the addition of an alkali or alkaline earth, then leaching the neutralized ore with a cyanide solution, and then passing the cyanide solution containing the precious metal through a sponge of zinc, substantially as set forth.

> JOHN STEWART MacARTHUR.
> ROBERT WARDROP FORREST.
> WILLIAM FORREST.

Witnesses:
  ROBERT JAMIESON MacKINLAY,
  CHARLES KEITH RITCHIE,
Both of 160 West George Street, Glasgow, Clerks at Law.

UNITED STATES PATENT OFFICE.

EDWARD D. KENDALL, of Brooklyn, New York.

COMPOSITION OF MATTER FOR THE EXTRACTION OF GOLD AND SILVER FROM ORES.

Specification forming part of Letters Patent No. 482,577, dated September 13, 1892.

(Application filed May 27, 1892. Serial No. 434,528. No specimens.)

To all whom it may concern:

Be it known that I, Edward D. Kendall, a citizen of the United States, residing at Brooklyn, in the county of Kings and State of New York, have invented a new and useful composition of matter to be used for the extraction of gold and silver from ores, so-called "tailings," and other matters containing one or both of these metals, of which composition the following is a specification.

My composition consists of the following ingredients, combined as hereinafter stated: water (hot or cold), potassium ferricyanide (or other soluble ferricyanide), and potassium cyanide (or other soluble cyanide). The best proportions of the two last mentioned constituents vary somewhat with the different ferricyanides and cyanides, and may be determined by calculation based on the molecular weights of the salts or the chemical equivalents of their elements, and also by considering that the purpose of my composition is to set free cyanogen to form—for example, when the potassium salts are used—the soluble double cyanide of gold or silver and potassium, and by applying my theory of the chemical reactions which occur, set forth in the following formula:

$$4 K_6 (C_3N_3)_4 Fe_2 + 16 KCN + 8 Au = 4 K_8 (C_3N_3)_4 Fe_2 + 8 K Au (CN)_2.$$

The indicated proportions are therefore, practically, by weight, two and one half parts of potassium ferricyanide and one part of potassium cyanide, and these proportions I have found satisfactory in practice; but they may be varied within wide limits without departing from my invention. The potassium ferricyanide, which is a product of the chemical action, facilitates the solution of the resulting cyanides, and after the separation of the precious metal from the menstruum by appropriate means, may be utilized by a process which I propose to make the subject of a separate application.

To prepare my composition, I dissolve the ferrocyanide in one portion of water and the cyanide in another portion, and mix the two solutions; or either salt, in solid form, may be added to the solution of the other. In dissolving the salts I do not always confine myself to a specific proportion of water. More or less water may be used. As a rule, the more concentrated the composition the more energetic its action, but the more costly. Except in treating substances very rich in gold or silver, my composition will always be used in a more or less dilute condition.

In using the herein-described composition, the gold- and silver-bearing minerals, tailings, and other matters, cold or while moderately heated, with or without prior chemical or mechanical treatment, should be placed in tanks, or troughs, or other receptacles made of any suitable material, as wood (if of wood, preferably lined with stoneware slabs), and thoroughly drenched, soaked, or impregnated with my composition, which is, after a time, to be drawn off and washed out, or displaced with water, in order that the contained precious metal may be separated by subsequent operations.

The composition may be used hot or cold. The effect of heat is to hasten the chemical and solvent action.

What I claim, and desire to secure by Letters Patent of the United States, is—

The before-described composition of matter—to be used for extracting gold and silver from minerals, tailings, and other matters containing one or both of these metals—consisting of water, one or more soluble ferricyanides, and one or more soluble cyanides, prepared and combined as herein stated.

EDWARD D. KENDALL.

Witnesses:
Edward M. McCook.
S. J. Storrs.

[Second Edition.]

No. 3,024.   A. D. 1892.

Date of Application, 16th Feb., 1892.

Complete Specification Left 16th Nov., 1892—Accepted 16th Feb., 1893.

PROVISIONAL SPECIFICATION.

IMPROVEMENTS IN PRECIPITATING AND COLLECTING METALS FROM SOLUTIONS CONTAINING THEM.

A communication by Bernard Charles Molloy, Member of Parliament, Barrister at Law, of the Middle Temple, London, temporarily residing in Johannesburg, South African Republic.

I, Alfred George Brookes, of 55 and 56 Chancery Lane, in the county of London, Chartered Patent Agent, do hereby declare the nature of this invention to be as follows:

This invention consists in a new method of precipitating and collecting gold and other metals from solutions containing these metals, such as bromide, chloride, and cyanide solutions.

In some cases these solutions are acid, but under the action of this process these solutions are rendered neutral and alkaline, so that the solutions are, or become, alkaline solutions of the metals which are soluble in alkaline solution.

The application of this process may be carried out in apparatus of many forms of construction and under many and differing conditions, depending on the character of the work required to be done.

The following example will, however, explain the nature of the process, and how it may be carried out:

Take a tray or tank constructed of wood, cement, or other suitable material and of a size as may be necessary.

Cover, or partially cover, the bottom of the tank with mercury. On this mercury will rest the solution from which the metals are to be precipitated. This mercury is then charged electrolytically with ammonium, sodium, potassium, or other alkaline metal, which is then amalgamated by the mercury. These alkaline metals, or the amalgams of these metals, when they come to the surface of the mercury and in contact with the water of the solution, decompose the water, the alkaline metal combining with the oxygen of the water to form an alkaline oxide, and so rendering the solution alkaline if not previously so, and the hydrogen of the decomposed water is at the same time evolved in a nascent state from the surface of the mercury in contact with and against the solution from which the metal in solution (say the gold) is to be precipitated and collected.

Under this action the metals in solution (such as gold) will be precipitated and absorbed by the mercury, from which it can be released in the ordinary manner by the action of heat.

In the above-described case the charging of the mercury with the alkaline metal has been effected electrolytically by the electrolysis of say an alkaline salt of the alkaline metal used in a porous vessel in contact with a mercury cathode or other convenient method.

Another though much less advantageous method is the mechanical addition to the mercury, of potassium or other alkaline metal or amalgam of the same, when the nascent hydrogen with an equivalent of the alkaline oxide will be produced.

These are some methods by which the process may be carried out, but there are others, as is evident, which may be employed.

In carrying out this process a solution of bromine, chlorine, or cyanogen may be used to dissolve the gold and other metals from their compounds, and then the process here described may be used to precipitate and collect the metals.

It is evident that generation of nascent hydrogen in contact with an alkaline solution of metals, soluble in such solution, may be obtained by other means, though not detailed here.

Dated the 16th day of February, 1892.

WM. BROOKES & SON,
55 and 56 Chancery Lane, London, Agents for the Applicant.

COMPLETE SPECIFICATION.

IMPROVEMENTS IN PRECIPITATING AND COLLECTING METALS FROM SOLUTIONS CONTAINING THEM.

A communication by Bernard Charles Molloy, Member of Parliament, Barrister at Law, of the Middle Temple, London, temporarily residing in Johannesburg, South African Republic.

I, Alfred George Brookes, of 55 and 56 Chancery Lane, in the county of London, Chartered Patent Agent, do hereby declare the nature of this invention and in what manner the same is to be performed, to be particularly described and ascertained in and by the following statement:

This invention consists in a new method of precipitating and collecting the gold and other metals from the solutions containing these metals, such as bromide, chloride, and cyanide solutions, when such are employed for dissolving out the precious metals from compounds containing the same.

In some cases these solutions are acid, but under the action of this process these solutions are rendered neutral and alkaline, so that the solutions are, or become, alkaline solutions of the metals, which are soluble in alkaline solutions.

The application of this process may be carried out in apparatus of many forms of construction, and under many and differing conditions, depending on the character of the work required to be done.

The following example of the treatment of a gold-bearing compound will, however explain the nature of the process, and how it may be carried out:

The crushed ore is treated, say, with a solution of cyanide of potassium, the quantity and saturation being in proportion to the work to be done. When the solvent solution is sufficiently charged, then the precipitation of the gold, the regeneration of this solvent, and the collection of the gold, is effected as follows:

Take a tray, or tank, constructed of wood, cement, or other suitable material, and of a size as may be necessary. Cover, or partially cover, the bottom of the tank with mercury.

On this mercury will rest, or pass over, the solution from which the metals are to be precipitated. This mercury is then charged electrolytically with ammonium, sodium, potassium, or other alkaline metal. These alkaline metals, when they come to the surface of the mercury and in contact with the water of the solution (now containing gold), decompose the water, the alkaline metal combining with the oxygen of the decomposed water to form an alkaline oxide, rendering the solution alkaline, if not previously so, and the hydrogen of the decomposed water is at the same time evolved in a nascent state from the surface of the mercury in contact with and against the solution from which the metal in solution (say the gold) is to be precipitated and collected. While the current continues to pass, the required nascent hydrogen and oxide will be produced and act on the solution containing the gold.

Under this action the metals in solution such as gold will be precipitated and absorbed by the mercury, from which it can be released in the ordinary manner by the action of heat.

In the above-described case, the charging of the mercury with the alkaline metal has been effected electrolytically by the electrolysis of, say, an alkaline salt of the alkaline metal used in a porous vessel in contact with the mercury cathode or other convenient method.

Another though less advantageous method is the mechanical addition to the mercury of potassium, or other alkaline metal, or amalgam of the same, when the nascent hydrogen, with an equivalent of the alkaline oxide, will be produced.

These are some methods by which the process may be carried out, but there are others, as is evident, which may be employed instead.

The action may be concisely described as follows:

The precious metal is dissolved out say by a solution of potassium cyanide. The solution is then brought into contact with mercury, charged, as described, with potassium. The potassium on coming into contact with the water of the solution decomposes it with the evolution of nascent hydrogen and the formation of the oxide of the alkaline metal. The hydrogen decomposes the solution of the new cyanide of gold—and sets the gold free, which is precipitated upon and collected by the mercury.

The metal of the alkaline oxide reacts upon the cyanogen compound, and so reforms or reproduces the cyanide of potassium. The original solution (of cyanide of potassium) is thus regenerated, and is then ready for re-use, thus effecting a great economy.

The following equations roughly represent the various reactions when cyanogen is the solvent:

$$H_2O + O + 4 KCN + 2 Au = 2 (KAu(CN)_2) + 2 KOH \quad 2 (KAu(CN)_2) + 2 H = 2 KCN + 2 CNH + 2 Au \quad 2 CNH + 2 KOH = 2 KCN + 2 H_2O.$$

In carrying out this process a suitable solution of a solvent for gold, such as bromine, or chlorine, or cyanogen, or their compounds, may be used to dissolve out the gold and other metals from their ores or compounds, and then the process more particularly described for precipitating the metals in such solution, and in some cases regenerating the solvent solutions and obtaining the metals.

It is evident that generation of nascent hydrogen in contact with an alkaline solution from which metals soluble in such solution are to be precipitated, may be obtained without the employment of a cathode of mercury, and using instead thereof another known cathode.

Although I have hereinbefore indicated some methods by which it may be worked, variations in detail may be effected without departing from the essential features of my invention.

Having now particularly described and ascertained the nature of my said invention, and in what manner the same is to be performed, I declare that what I claim is—

1. The method of obtaining gold and other metals from solutions which have been employed in dissolving out such precious metals from ores, or compounds, containing the same, substantially in the manner and for the purpose hereinbefore set forth.

2. The described method of precipitating and collecting gold and other metals from solutions, such as referred to, containing them, by the action of the alkaline metals, in the manner hereinbefore set forth.

3. The extraction of gold or other metals from ores, or other compounds, by solutions of cyanogen, or its compounds, the precipitation of gold or other metals from such solutions, the regeneration of such solutions, and the collection of the gold or other metal, all substantially as and for the purpose set forth.

4. The described method of precipitating gold and other metals from alkaline solutions, such as indicated, containing them, by the action of nascent hydrogen, in the manner hereinbefore set forth.

5. The precipitation of gold, and other precious metals from solutions, such as indicated, containing them, by means of the alkaline metals, or amalgams of the same, obtained electrolytically, substantially as and for the purpose set forth.

6. The separation and collection of gold or other precious metals from their ores or compounds by means of a suitable solvent for said metals, an electrolyte of the alkaline metals, a current of electricity, a mercury cathode, all substantially as set forth.

7. The employment of bromine, chlorine, iodine, and cyanogen, or their compounds, as solvents for gold or other metals, in combination with the above-described process for effecting the precipitation and collection of the gold and other metals in solution therein, substantially as set forth.

Dated the 16th day of November, 1892.

WM. BROOKES & SON,
55 and 56 Chancery Lane, London, Agents for the Applicant.

[Second Edition.]

No. 12,641.   A. D. 1892.

Date of Application, 8th July, 1892.

Complete Specification Left 10th April, 1893—Accepted 8th July, 1893.

PROVISIONAL SPECIFICATION.

IMPROVEMENTS IN THE EXTRACTION OF GOLD AND SILVER FROM ORES OR COMPOUNDS CONTAINING THE SAME, AND IN APPARATUS APPLICABLE FOR USE IN THE TREATMENT OF SUCH MATERIALS BY MEANS OF SOLVENTS.

I, John Cuninghame Montgomerie, of the "Water of Ayr" and "Tam O'Shanter" Hone Works, Dalmore, Stair, in the county of Ayr, manufacturer, do hereby declare the nature of this invention to be as follows:

This invention relates to the treatment of auriferous and argentiferous ores or compounds, for the purpose of separating and collecting the gold and silver contained therein, by means of solvent agents—as, for example, cyanide of potassium—and to apparatus applicable for use in processes of this description.

According to the method usually employed in the recovery of gold and silver by means of cyanide of potassium, the ore or other material having been reduced to a finely-triturated state is placed, along with the solvent, in a barrel or other vessel and is there subjected to agitation. After the lapse of a few hours the contents of the barrel are removed to a filter, where the liquid portion of the charge (containing the precious metals in solution) is separated from the ore. The latter is further washed for the removal of any gold or silver remaining (in solution) therewith. The cyanide solution of gold and silver (as also the wash-water) is then treated for the recovery of the precious metals by precipitation.

When a cyanide solvent is employed as hereinbefore described, the proportion of cyanide is necessarily considerably in excess of that required for chemical combination with the gold and silver present in the ore. This excess remains with the liquor after the precious metals have been precipitated therefrom, and is either run to waste or is subjected to a separate process for the recovery of the cyanide.

My improvement in the process of extraction by the method hereinbefore referred to consists in applying the cyanide solution of gold and silver, after having been separated from the ore by filtration, to a subsequent charge or to subsequent charges of fresh ore prior to treating the solution for the separation of the precious metals by precipitation, care being taken that the original quantity of water is made good. With this object the requisite quantity of water is preferably added as soon as the surface of the ore contained in the filter presents a dry appearance, the added water displacing the liquid remaining in the ore and permitting the same (which is highly charged with the solvent and with the precious metals in solution) to be discharged. The solution is then tested for cyanide of potassium (or such other solvent agent as may be employed), and the deficiency supplied by the addition of a suitable quantity of cyanide of potassium (or other solvent agent), thereby restoring the solvent solution to its original strength before adding the same to the fresh charge of ore. Before being added to the fresh charge of ore, the solution is made slightly alkaline by the addition of an alkali, preferably caustic soda.

Where cyanide is employed, it is necessary for the ore to be thoroughly neutralized before treatment; and in some cases it is advantageous to have it slightly alkaline, especially where oxygen is used under pressure. The tailings are then further washed to remove the last trace of gold and silver remaining in solution, and the resultant wash-water is treated in the usual way for the recovery of the precious metals contained therein.

By this mode of procedure considerable economy is effected, both in the quantity of cyanide or other solvent used and in the cost of working, the quantity of liquid subjected to treatment for the recovery of the gold and silver by precipitation being at the same time greatly reduced.

My invention relates secondly to the construction of the barrel or other vessel in which the ore is subjected to the action of the solvent.

If this barrel or vessel be formed of metal, its internal surface is rapidly acted upon by cyanide of potassium or other solvent of the precious metals; and if a lining of wood or similar material be employed, the latter is incapable of withstanding the chemical action of the solvent and the abrasive action of the ore for any length of time.

With a view to overcoming these difficulties, I line the barrel or vessel with tiles or segments composed of glass or glazed porcelain or similar solvent- and acid-resisting material, the same being set in cement adapted to withstand the chemical action of the cyanide or other solvent employed.

My invention relates thirdly to the construction of the filter or leaching vat employed for separating the ore from the cyanide or other solution of gold and silver, or from the wash-water.

A filter constructed according to my improved method comprises an upper vessel for the reception of the mixture of ore and solvent, and a lower vessel in which the solution is received after passing through the filter-bed. The latter is formed of filter cloth carried on wire gauze coated with an acid-proof enamel and supported on wooden laths. The upper vessel is attached to the lower vessel by means of bolts, and is so arranged that the bottom edge of the former rests upon the periphery of the filter cloth and grips the same in a recess formed in the upper edge of the lower vessel, thereby securing a water-tight joint between the two vessels and at the same time holding the filter cloth securely in position.

The filter or leaching vat may be lined with segments or tiles in the manner herein before described with reference to the barrel or other vessel in which the ore is subjected to the action of the solvent.

Dated the 6th day of July, 1892.

G. G. M. HARDINGHAM,
191 Fleet Street, London, E.C., Chartered Patent Agent.

COMPLETE SPECIFICATION.

IMPROVEMENTS IN THE EXTRACTION OF GOLD AND SILVER FROM ORES OR COMPOUNDS CONTAINING THE SAME, AND IN APPARATUS APPLICABLE FOR USE IN THE TREATMENT OF SUCH MATERIALS BY MEANS OF SOLVENTS.

I, John Cuninghame Montgomerie, of the "Water of Ayr" and "Tam O'Shanter" Hone Works, Dalmore, Stair, in the county of Ayr, Scotland, manufacturer, do hereby declare the nature of this invention, and in what manner the same is to be performed, to be particularly described and ascertained in and by the following statement:

This invention relates to the treatment of auriferous and argentiferous ores or compounds, for the purpose of separating and collecting the gold and silver contained therein, by means of solvent agents—as, for example, cyanide of potassium—and to apparatus applicable for use in processes of this description.

According to a method commonly employed in the recovery of gold and silver by means of cyanide of potassium, the ore or other material having been reduced to a finely-triturated state is placed, along with the solvent, in a barrel or other vessel, and is there subjected to agitation. After the lapse of a few hours, the contents of the barrel are removed to a filter, where the liquid portion of the charge (containing the precious metals in solution) is separated from the ore. The latter is further washed for the removal of any gold or silver remaining (in solution) therewith. The cyanide solution of gold and silver, as also the wash-water, is then treated for the recovery of the precious metals by precipitation in a zinc filter or percolator.

When a cyanide solvent is employed as hereinbefore described, a certain portion thereof is taken up by base metals and other impurities invariably present in greater or less proportions in the ore. The solvent is also contaminated by the zinc dissolved whilst the mixture of ore and solvent is under treatment in the zinc percolator; both of these causes resulting in a considerable waste of the cyanide, and in its contamination with deleterious matter.

My improvement in the process of extraction by means of the kind hereinbefore referred to, consists in adding sodium oxide (caustic soda), or other suitable oxide of the alkalies, to the cyanide solution before (or whilst) mixing the same with the ore, thereupon agitating or otherwise treating the resultant mass for the time requisite for enabling the gold and silver to be dissolved by such a solution, then discharging the same into a filter and drawing off the original quantity of water employed, the same being highly charged with the unconsumed cyanide and sodium oxide, and with the precious metals in solution.

By the employment of sodium oxide in the manner hereinbefore described, particularly where the alkali is in excess, I have found that the proportion of cyanide or other solvent may be considerably reduced and an important economy in the cost of working effected.

In carrying out this stage of the process, a sufficient quantity of water is added to the surface of the ore in the filter as soon as it becomes dry, the added water displacing the liquid remaining in the ore and permitting the latter to be discharged. The liquid obtained is then tested for cyanide of potassium and sodium oxide, and the deficiency supplied by the addition of the necessary quantity of these agents so as to restore the solvent solution to its original character and strength. This solution is now applied to a fresh charge of ore, and the same operation is repeated with successive charges till it is found necessary to discharge the solution with a view to precipitating the gold and silver in the usual manner. Experiment alone can determine the quantity of solvent and of sodium oxide appropriate, and the period of time requisite to insure the greatest extraction of the precious metals with the least consumption of the solvent, as these will vary according to the nature of the ore operated upon. (It may be mentioned by

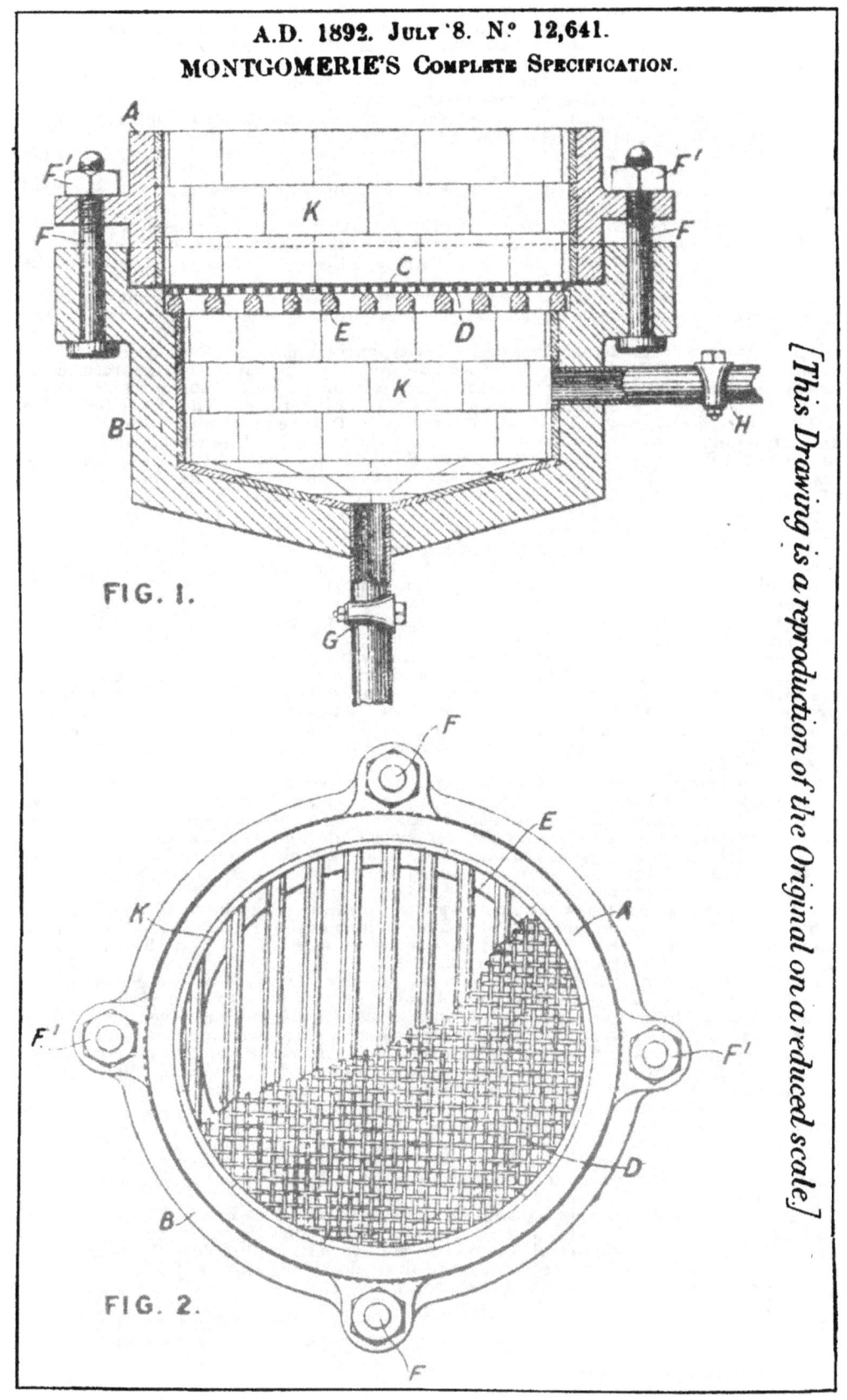

A.D. 1892. JULY 8. N.° 12,641.

MONTGOMERIE'S COMPLETE SPECIFICATION.

FIG. 1.

FIG. 2.

[This Drawing is a reproduction of the Original on a reduced scale.]

way of illustration that for an ore containing about 4 ozs. of gold and 12 ozs. of silver per ton, 12 lbs. of cyanide of potassium and 3 lbs. of sodium oxide would be suitable.) The tailings are then further washed to remove the last trace of gold and silver remaining in solution, and the resultant wash-water is treated in the usual way for the recovery of the precious metals contained therein.

By this mode of procedure, the quantity of liquid subjected to treatment for the recovery of the gold and silver by precipitation is greatly reduced.

My invention relates secondly to the construction of the barrel or other vessel in which the ore is subjected to the action of the solvent.

If this barrel or vessel be formed of metal, its internal surface is rapidly acted upon by cyanide of potassium or other solvent of the precious metals; and if a lining of wood or similar material be employed, the latter is incapable of withstanding the chemical action of the solvent and the abrasive action of the ore for any length of time.

With a view to overcoming these difficulties, I line the barrel or vessel with tiles or segments composed of glass or glazed porcelain, or similar solvent- and acid-resisting material, the same being set in cement adapted to withstand the chemical action of the cyanide or other solvent employed.

My invention relates thirdly to the construction of the filter or leaching vat employed for separating the ore from the cyanide or other solution of gold and silver, or from the wash-water.

Apparatus constructed according to my invention is illustrated in the accompanying drawings, whereof Fig. 1 is a vertical section, and Fig. 2 a plan. The apparatus comprises an upper vessel A for the reception of the mixture of ore and solvent, and a lower vessel B in which the solution is received after passing through the filter-bed. The latter is formed of filter-cloth C carried on wire gauze D coated with an acid-proof enamel and supported on wooden laths E on the top of the vessel B. The vessel A is attached to the vessel B by means of bolts F and nuts F¹, and is so arranged that its bottom edge rests upon the circular margin of the filter-cloth C, which it presses against the bottom of a recess or socket formed in the upper edge of the vessel B, thereby securing a water-tight joint between the two vessels and at the same time holding the filter-cloth securely in position. G is a draw-off cock; H being the exhaust cock. The vessels A and B are lined with segments or tiles K, composed of glass, glazed porcelain, or similar solvent- and acid-resisting material, set in cement adapted to withstand the chemical action of the cyanide or other solvent employed. When a barrel is used, it may be lined with segments or tiles set in a similar manner.

Having now particularly described and ascertained the nature of this invention, and in what manner the same is to be performed, I claim—

1. The improved process of extracting gold and silver from ores or compounds containing the same, substantially as herein described; the same consisting in mixing the ore with a solution of cyanide of potassium or other cyanide solvent rendered alkaline by the addition of sodium oxide or an equivalent alkaline oxide, filtering or otherwise separating the liquid (containing the gold and silver in solution) from the ore, and treating the former, by precipitation or other known mode, for the recovery of the precious metals.

2. In the extraction of the precious metals by a solvent process of the general character herein referred to, applying the solvent solution, after separation from the first charge of ore, to a subsequent charge or successively to subsequent charges of fresh ore, the solution being fortified at each operation by the addition of a suitable quantity of the chemical agents employed, and ultimately treating the liquid (consisting of a more or less saturated solution of gold and silver) by any known means for the separation and recovery of the precious metals.

3. In the process of extracting gold and silver by means of cyanide of potassium or other cyanide solvent, the addition of sodium oxide or other suitable alkaline oxide to the solvent, either prior to or during its admixture with the ore, for the purpose of economizing the solvent and expediting its action.

4. In the extraction of the precious metals by a solvent process of the general character herein referred to, discharging the solvent remaining with the ore after filtration by adding water to the surface of the ore and thereby displacing the solvent containing the precious metals in solution, substantially as herein described.

5. In apparatus adapted for use in the treatment of ores or compounds containing gold or silver, a barrel, filter, or leaching vessel such as A or B lined with tiles K set in an acid- or solvent-resisting cement, substantially as herein described.

6. The herein described apparatus for use in the treatment of ores or compounds containing gold and silver by means of solvents, the same comprising an upper vessel A for the reception of the ore and solvent, a lower vessel B in which the solution is received, a filter cloth C held between the lower part of the vessel A and a socket in the upper part of the vessel B, wire gauze D on which the filter cloth lies, and bars E for supporting the wire gauze.

7. The herein described apparatus for use in the treatment of ores or compounds containing gold and silver by means of solvents, the same comprising an upper vessel A lined with tiles K, a lower vessel B also lined with tiles K, a filter cloth C held between the vessels A and B, wire gauze D under the filter cloth, bars E for supporting the wire gauze, a draw-off cock G, and an exhaust cock H.

Dated the 8th day of April, 1893.

G. G. M. HARDINGHAM,
191 Fleet Street, London, E.C., Chartered Patent Agent.

UNITED STATES PATENT OFFICE.

ALEXIS JANIN AND CHARLES W. MERRILL, OF SAN FRANCISCO, CALIFORNIA.

PROCESS OF LEACHING ORES WITH SOLUTIONS OF ALKALINE CYANIDES.

Specification forming part of Letters Patent No. 515,148, dated February 20, 1894.

(Application filed June 12, 1893. Series No. 477,338. No specimens.)

To all whom it may concern:

Be it known that we, Alexis Janin and Charles W. Merrill, citizens of the United States residing in the city and county of San Francisco, State of California, have invented an Improvement in Processes of Leaching Ores with Solutions of Alkaline Cyanides; and we hereby declare the following to be a full, clear, and exact description of the same:

Our invention relates to an improvement in the art of leaching ores with solutions of alkaline cyanides, and consists in, first, precipitating and separating, in the form of silver sulphide, by means of an alkaline sulphide or of sulphuretted hydrogen gas, all or the greater portion of the silver dissolved from the ore by such solutions, and then precipitating in the metallic state, by means of metallic zinc, the gold contained in the same solution, together with any silver which has escaped precipitation as a sulphide.

In the usual method of leaching ores with a solution of potassium cyanide, the gold and silver extracted are both precipitated from the solution in the metallic state, with metallic zinc. When much silver is present, this method involves a large consumption of zinc and consequent contamination of the cyanide solution by the zinc dissolved, and unless the contact between the zinc and the silver-bearing solution be greatly prolonged, the precipitation of the silver is imperfect. Furthermore, the potassium cyanide which combines with the zinc dissolved is practically lost. When the silver is precipitated from its solution in potassium cyanide by means of an alkaline sulphide an alkaline cyanide is regenerated, which is again available for leaching. If sulphuretted hydrogen gas be used to precipitate the silver there is also formed free hydrocyanic acid, but if the solution of potassium cyanide contains free alkali, or if such be added to the solution, no free hydrocyanic acid will escape, either because the sulphuretted hydrogen gas first combines with the alkali, to form a sulphide which precipitates the silver in the manner described, or because any hydrocyanic acid generated will also combine with the free alkali to form an alkaline cyanide.

We have found that, whereas, silver is not precipitated at all, or only very imperfectly, from strong solutions of potassium cyanide, by means of the agents hereinafter mentioned, yet when the silver-bearing solution contains only about one and one half per cent, or less, of pure potassium cyanide (KCN) or its equivalent, then the silver can be thoroughly precipitated by means of the sulphides of sodium, potassium, or ammonium, or by sulphuretted hydrogen gas, and the precipitation of the silver becomes more imperfect as the strength of the solution in potassium cyanide is increased. Therefore when leaching silver-bearing ores we employ solutions containing, at the most, two per cent of potassium cyanide or its equivalent. As a precipitating agent we employ preferably a solution of sodium sulphide, approaching, as nearly as practicable, to the composition of a monosulphide, in order to avoid, as much as possible, the separation of free sulphur in precipitating the silver.

In practice we leach ores containing both gold and silver, with a solution of potassium cyanide containing not more than two per cent of KCN or its equivalent, or as much weaker as is consistent with a thorough extraction. The solution, after passing through the ore, is run into precipitating vats, where a solution of sodium sulphide is added in sufficient quantity to convert the silver present into sulphide of silver, or in a little less than that amount, in order to avoid the possibility of any excess of the precipitating agent remaining in the solution which might be prejudicial in its further use. The precipitate of silver sulphide is allowed to settle, the supernatant solution of potassium cyanide is then drawn off, and the gold, together with any silver remaining in solution, is precipitated by means of metallic zinc.

Having thus described our invention, what we claim as new, and desire to secure by Letters Patent, is—

The improvement in the art of leaching ores with solutions of alkaline cyanides, which consists in first leaching the ore with such solutions, then adding to the solution an agent which will precipitate the silver present as a sulphide, and then precipitating the gold in the solution with metallic zinc, substantially as herein described.

In witness whereof we have hereunto set our hands.

ALEXIS JANIN.
CHARLES W. MERRILL.

Witnesses:
  S. H. NOURSE.
  WM. F. BOOTH.

NEW ZEALAND PATENT OFFICE.

——

JOHN STEWART MacARTHUR AND CHARLES JAMES ELLIS.

APPLICATION FOR LETTERS PATENT FOR IMPROVEMENTS IN EXTRACTING GOLD AND SILVER FROM ORES AND THE LIKE.

——

We, John Stewart MacArthur, Managing Director of the Cassel Gold Extracting Company (Limited), and Charles James Ellis, technical chemist to the said company, both of 157 West George Street, Glasgow, in the county of Lanark, North Britain, do declare the nature of our invention for " Improvements in extracting gold and silver from ores and the like," and in what manner the same is to be performed, to be particularly described and ascertained in and by the following statement:

Our said invention relates to what is known as the "MacArthur-Forrest process" for extracting gold and silver from ores and the like by means of cyanides, and has for its object to increase the efficiency and economy of that process in cases in which, from the nature of the ores treated, or other circumstances, it is found that in the solution of cyanide as heretofore used there is formed, or becomes present, a sulphide soluble therein which retards and objectionably affects the action of the cyanide on the precious metals.

Our invention consists in removing or rendering inert such sulphide by adding to the solution of cyanide, or to the ore, or to the mixture of ore and cyanide solution, a suitable salt or compound of a metal which will form with the sulphur of the sulphide a sulphide which is practically insoluble or inert in the cyanide solution, or which will materially diminish the objectionable action.

In carrying out our invention, we may use any one or more of various metallic salts or compounds, of which the following may be mentioned, by way of example, preference being given to them in the order in which they are noted, namely : Salts or compounds of lead, such as plumbates, carbonate acetate, or sulphate of lead; or salts or compounds of other metals, such as sulphate or chloride of manganese, zincates, oxide, or chloride of mercury, and ferric hydrate, or oxide. The proportion to be used in any case will depend on the proportion of soluble sulphide which has to be dealt with in the cyanide solution applied to the particular ore, and is easily and most conveniently ascertained by trials of a few samples in each case. In the case of some ores containing sulphur, we find the addition of salts or compounds, as and for the purpose hereinbefore referred to, and especially those of lead and mercury, increases the percentage of precious metals obtained.

Having now particularly described our said invention, and in what manner the same is to be performed, we declare what we claim is—

1. In the MacArthur-Forrest process for extracting gold and silver from ores and the like, the addition to the cyanide solution, or to the ore, or to the mixture of ore and cyanide, of salts or compounds of lead, substantially as and for the purposes hereinbefore described.

2. In the MacArthur-Forrest process for extracting gold and silver from ores and the like, the addition to the cyanide solution, or to the ore, or to the mixture of ore and cyanide, of any one or more of the metallic salts or compounds hereinbefore indicated, and capable of forming insoluble sulphides, as and for the purposes hereinbefore described.

Dated June 29, 1893.

JOHN STEWART MacARTHUR.
CHARLES JAMES ELLIS.

NEW ZEALAND PATENT OFFICE.

CARL MOLDENHAUER.

APPLICATION FOR LETTERS PATENT FOR IMPROVEMENTS IN RECOVERING GOLD AND OTHER PRECIOUS METALS FROM THEIR ORES.

I, Carl Moldenhauer, of Frankfort-on-Main, in the Empire of Germany, do hereby declare that the nature of my invention for improvements in recovering gold and other precious metals from their ores, and the manner in which the same is to be used, are particularly described and ascertained in and by the following statement:

This invention relates to extracting gold and other precious metals from their ores by means of a solution of cyanide of an alkali or an alkaline earth, and has for its object to render the process more expeditious and considerably cheaper than heretofore.

The invention consists, firstly, in adding to the cyanide solution an artificial oxidizing agent; and secondly, in precipitating the extracted precious metal out of its cyanide solution by means of aluminium or alloys or amalgam thereof.

As to the first part of my invention, I have found that the dissolving action of the cyanide solution on the precious metal is highly expedited, and much cyanide is saved, if an artificial oxidizing agent is added to the said cyanide solution. As such an artificial oxidizing agent, I use, by preference, ferricyanide of potassium, or another ferricyanogen salt of an alkali or of an alkaline earth. In either case the ferricyanogen salt is preferably employed in alkaline solution. The result of this addition of an artificial oxidizing agent is that the dissolving action of the solvent is rendered more energetic, and consequently a considerably smaller quantity of the solvent is required. Thus, by the addition of ferricyanide of potassium or other ferricyanogen salt in alkaline solution, as much as 80 per cent of the potassium may be saved. The proportions preferred are from one half to two parts of ferricyanide to one part of cyanide.

It may be remarked that the ferricyanide of potassium alone will not dissolve the gold, and does not, therefore, come under the category of the solvent heretofore employed in processes of extraction. It does not, therefore, render unnecessary the employment of the simple cyanide as a solvent, but only reduces the amount required, owing to the capacity of the ferricyanide to act as an oxidizing agent; consequently the cyanogen of the ferricyanide is not used to form the gold cyanide compound. I may also employ permanganate of potash as the oxidizing agent instead of the ferricyanide. The said permanganate of potash is also added in solution and in the same proportions as before, namely, from one half to two parts of permanganate to one part of cyanide.

In lieu of permanganate of potash, any other suitable oxidizing agent can be used in carrying out my invention in practice, the invention not being restricted to the use of any special oxidizing agent, but includes the use of an agent that exerts an oxidizing action in the cyanide solution. The process can be carried out in a ball-mill lined with porcelain, wood, or other substance unattackable by the chemicals employed.

I may also use the cyanide solution and the oxidizing agent in combination with a preliminary treatment of the ore with any acid or salt that renders the precious metal better adapted to the subsequent treatment of the cyanide solution.

The second part of the process consists in precipitating the dissolved gold or precious metal out of its cyanide solution by means of aluminium alloy or aluminium amalgam; but this can also be applied with advantage to sulphurized solutions containing free alkali—that is to say, solutions which contain gold in the form of sulphuret, or hyposulphide of gold.

Zinc has heretofore been employed in practice by preference in precipitating gold from the cyanide solutions obtained by leaching auriferous ores. The employment of zinc for this purpose is found, however, to be attended with serious disadvantages. Now, I have discovered that aluminium can be employed for this purpose in place of zinc, without the disadvantages attending the use of the latter.

Whilst zinc forms a combination with the bound or free compound of cyanogen and alkali contained in the cyanide solution, aluminium separates the gold very quickly from the cyanogen solution without entering into combination with the cyanogen, but simply reacting with the caustic alkali which is present at the same time. By the action of aluminium the cyanide of potassium employed for leaching the gold out of its ore is regenerated, which is not the case when zinc is employed. But the zinc does not confine itself to entering into combination with the cyanogen of the cyanogen compounds of the gold, but also acts upon the free cyanide of potassium contained in the solution, so that a great part of the latter is consumed; but this is not the case when aluminium is employed.

These results are of the greatest importance when the solutions separated from the gold is to be employed in subsequent gold-extracting operations, as the whole of the cyanogen in the regenerated and liberated cyanide of potassium is enabled to renew its action; but the lyes resulting from the employment of zinc cannot be employed with the same advantage in subsequent operations for the extraction of gold. Numerous attempts have been made to regenerate the zinc, but are found to be inconvenient and costly. It is consequently evident that an important saving in cyanide of potassium is obtained by the employment of aluminium.

Aluminium acts in a like manner in a sulphurized alkaline solution—that is to say, in a solution containing the gold in the form of sulphuret of gold or hyposulphide of gold. It does not enter into combination with the sulphur in a solution of this description. This great and important advantage attending to the employment of aluminium, aluminium alloys, or aluminium amalgam, is combined with other advantages, as follows:

Aluminium is far less subject to oxidation than is zinc, so that it can be sent from its place of production in the form in which it is to be used for the precipitation, whereas when zinc is employed it is considered to be an important advantage to reduce it to the required form at the place where it is employed, and immediately before using it. For the same reason, the repeated employment of the aluminium is admissible for continuous precipitation.

Finally, the quantity of aluminium required for precipitating the same quantity of precious metal is about four times less than the amount of zinc required to produce the same effect.

I am aware that attempts have been made to employ aluminium for precipitating precious metals from acid or neutral solutions, but in this case it offers no advantages as compared with zinc and iron.

On the other hand, the practical precipitation of precious metals from alkaline cyanide solutions or sulphurized solutions by means of aluminium was not known, neither was it known that by the employment of the same in the presence of free alkali it was possible to obtain the important advantages hereinbefore set forth.

Of course, instead of pure aluminium, alloys or an amalgam thereof can be used with a like advantage; furthermore, I do not confine myself to the use of the aluminium, its alloys or amalgam, in any special form, as it may be used in any suitable form without departing from my invention.

Having particularly described and ascertained the nature of my said invention, and in what manner the same is to be performed, I declare that what I claim is—

1. Extracting gold and other precious metals from their ores by subjecting the ores to the dissolving action of a cyanide of an alkali or an alkaline earth in the presence of an oxidizing agent, substantially as and for the purpose hereinbefore set forth.

2. Extracting gold from its ores by subjecting the ores first to the action of an acid and subsequently to the dissolving action of cyanide of an alkali or an alkaline earth in the presence of an oxidizing agent, substantially as and for the purpose hereinbefore set forth.

3. Extracting gold from its ores by subjecting the ores to the dissolving action of cyanide of an alkali or an alkaline earth in a ball-mill, substantially as and for the purpose hereinbefore set forth.

4. Precipitating gold or other precious metals out of their solutions by means of aluminium, aluminium alloys, or aluminium amalgam, in the presence of a free alkali, substantially as hereinbefore described.

Dated August 31, 1893.

                                                    CARL MOLDENHAUER.

NEW ZEALAND PATENT OFFICE.

CARL MARIA PIELSTICKER.

APPLICATION FOR LETTERS PATENT FOR IMPROVEMENTS IN THE EXTRACTION OF GOLD AND SILVER FROM ORES.

I, Carl Maria Pielsticker, of No. 43 Connaught Road, Harlesden, in the county of Middlesex, England, engineer, do hereby declare the nature of my invention for Improvements in the Extraction of Gold and Silver from Ores, and in what manner the same is to be performed, to be particularly described and ascertained in and by the following statement:

My invention has for its object the extraction of gold and silver, particularly from sulphide, and from such ores in which the precious metals exist in a state of extremely fine division, and it consists essentially in treating the powdered ore with a solution of cyanide of potassium or a cyanide or cyanogen-yielding substance in conjunction with an electric current, continuous circulation of the solvent, continuous precipitation of the dissolved precious metals by electrolysis, and continuous regeneration of the solvent.

In carrying out my invention, I employ a tank, marked A on the accompanying drawing, which I call the ore-tank, in which the ore is subjected to the treatment with cyanide of potassium in conjunction with an electric current. About 3 inches from the bottom I place a perforated plate, H, preferably of iron or carbon, covered with a porous material, such as cocoanut matting. The plate H serves as anode, and is insoluble, or practically so, in cyanide of potassium.

If made of iron, I prefer a highly carburetted iron, or ore containing a high percentage of silicum. Near the top of the ore-tank I place a second perforated plate, G, which serves as cathode. Both these plates are connected by means of insulated wires, e—e, with a dynamo, D, or other source of electricity.

The ore-tank A is connected near its top by means of a pipe with a second tank, B, containing a number of baffle plates, K, or their equivalent, which are destined to arrest any suspended matter flowing over with the solution from the ore-tank, and which otherwise would greatly interfere with the precipitation of the dissolved precious metals in the following tank, C, connected with the tank B near the top by means of a pipe. The precipitating-tank C contains one or more pairs of electrodes, M and N, connected with the dynamo D, or other source of electricity, by means of the insulated wires, g and g¹, of which the anode preferably consists of carbon or other material, insoluble, or practically so, in cyanide of potassium. A pump P is connected with the ore-tank A under the anode H on the one hand, and with the top of the depositing-tank C on the other hand, enabling me to maintain a circulation of the solvent through the set of tanks.

In operating my invention, I fill the ore-tank A between the electrodes H and G with powdered ore, and admit into the three tanks A, B, and C, a solution of cyanide of potassium, filling them above the level of the pipes which connect one tank with the other. The strength of the cyanide solution may vary, care being taken to have sufficient cyanogen present to bring the gold and silver in the ore into the solution, the amount of which has previously been ascertained by assay. I connect the electrodes H and G in the ore-tank A, and M and N in the depositing-tank C, with the dynamo D, or other source of electricity, and force the cyanide solution from below through the ore in the tank A.

The solution pregnant with dissolved precious metals overflows into the settling-tank B, where it clears itself from suspended matter, and becomes thus fit to part with the dissolved precious metals on overflowing into the depositing-tank C, where the latter are precipitated on the cathode, and from which they are recovered by amalgamation or otherwise. The cyanide solution, freed from dissolved metals, and therefore in a better condition to dissolve more metal than when loaded with metal in solution, is pumped from the depositing-tank C, again through the ore in the tank A, where it dissolves a fresh proportion of precious metals, and so on, continuously, until the precious metals contained in the ore under treatment are dissolved.

In this manner my process becomes a continuous one, of dissolving the precious metals from the ore, preparing the solution pregnant with dissolved metals for electrolysis by separating continuously the suspended matter therefrom, precipitating continuously the dissolved metals by electrolysis, and regenerating continuously the solvent for further action on the undissolved precious metals still contained in the ore. Very little of the precious metals are precipitated on the cathode G in the ore tank, as the amount of the suspended matter present in the solution interferes with precipitation in this tank.

The electric current in the depositing-tank must be of low tension, and so regulated as to be of just sufficient strength to deposit the gold and silver without also decomposing the cyanide of potassium; the gold and silver, being more readily precipitated from their double salts of cyanide of gold (or silver) and potassium than the cyanogen, is set free from the simple salt of cyanide of potassium so long as the current of electricity is sufficiently low in tension, and so long as there are metals present in the solution.

The original solution can therefore be used over and over again for a long time, and only the loss made good occasionally.

In practice I find that an electro-motive force of about one volt, and an intensity of about ten ampères per square meter of surface of cathode, is well adapted for depositing the gold and silver in the tank C. I may find it desirable to employ a current of electricity of greater potential in the ore-tank A and of lesser potential in the depositing-tank C.

The great advantage in treating ores with cyanide of potassium in conjunction with an electric current lies in the fact that the precious metals are attacked by the cyanide solution more energetically in conjunction with a current of electricity than without one; further, when the dissolved precious metals are precipitated by means of an electrical current and an insoluble anode very little cyanide and no metal is consumed, as is the case when, for instance, zinc is used as a precipitant, when not only zinc is consumed, but also cyanide of potassium in the formation of a double salt of cyanide of zinc and potassium. Moreover, serious losses in gold and silver are occasioned in the recovery of the precious metals from the zinc slimes, whereas nothing can be simpler than their recovery from the cathode by amalgamation. Again, the precipitation of gold and silver is greatly accelerated by the electric current.

When these metals are precipitated by zinc without a current of electricity, the latter goes into solution as a double salt of cyanide of zinc and potassium, but the amount of zinc which is converted into cyanide of zinc is directly proportionate to the time during which it is in contact with the cyanide solution. Therefore, the more time is consumed in precipitating the gold and silver, the more cyanide and the more zinc will be wasted.

The cyanide process is most advantageously employed on ores in which either the gold is so finely divided in a free state that it is difficult to retain it by older methods, or for sulphide ores. Free gold is certainly more quickly dissolved by cyanide of potassium in conjunction with an electrical current than without one. As regards pyritic ores, if they are simply iron pyrites, as they are in a great number of cases, a cyanide of potassium solution, whatever its strength may be, has as little action on them when used in conjunction with an electric current of the strength I use as without one, only the gold and silver in the ore are more quickly dissolved in conjunction with an electrical current than without one.

If the ores contain sulphides, oxides, or carbonates, for instance, of copper and zinc, these are as easily dissolved by a cyanide of potassium solution, whether employed by itself or in conjunction with an electric current such as I use. Such ores, however, I prefer to treat first with, say, a 5 per cent sulphuric acid or other acid solution in water, or a strong solution of sulphurous acid in water in sufficient quantity to dissolve such metals, then leach with water, and then treat with the cyanide solution in conjunction with the electric current.

I would have it understood that I do not limit myself to the precise details herein set forth and illustrated on the drawing—for example, the number, nature, and position of electrodes, of the sources of electricity, and of the number, shape, and position of the tanks; all may be varied while retaining the construction and combinations for the proper carrying out of my process of extraction of gold and silver from their ores; further, I am aware that cyanide of potassium has been used in conjunction with an electric current for like purposes, and I make no broad claim thereto.

Having now particularly described and ascertained the nature of my said invention, and the manner in which the same is to be performed, I declare that what I claim is—

1. The process of separating gold and silver from their ores, which consists in treating the powdered ore with a solution of cyanide of potassium in conjunction with an electric current, depositing the precious metals constantly by means of a current of electricity of low tension and electrodes, of which the positive one is insoluble in cyanide of potassium, and bringing the cyanide of potassium solution thus freed from dissolved metals constantly again into contact with the ore, whereby I obtain a continuous process of extraction and precipitation, all substantially as herein described.

2. In the process of separating gold and silver from their ores by means of a solution of cyanide of potassium in conjunction with an electric current, bringing the cyanide of potassium solution freed from dissolved metals continuously into contact with the ore, substantially as described.

3. In the above-described process of separating gold and silver from ores by means of a solution of cyanide of potassium in conjunction with an electric current, depositing the dissolved metals by means of electrodes contained in depositing-tank or tanks, an electric current being passed through the ore-tank and depositing-tank, substantially as set forth.

4. In the above-described process of separating gold and silver from their ores by means of a solution of cyanide of potassium in conjunction with an electric current, treating the ore with an acid in combination with a subsequent treatment of cyanide of potassium in conjunction with an electric current and continuous circulation of the solution, substantially as described.

5. In the above-described process of separating gold and silver from their ores by means of a solution of cyanide of potassium in conjunction with an electric current, subjecting the ore and solution in the ore-tank to an electric current of greater potential, and depositing the dissolved metal in a depositing-tank by an electric current of lesser potential, substantially as described.

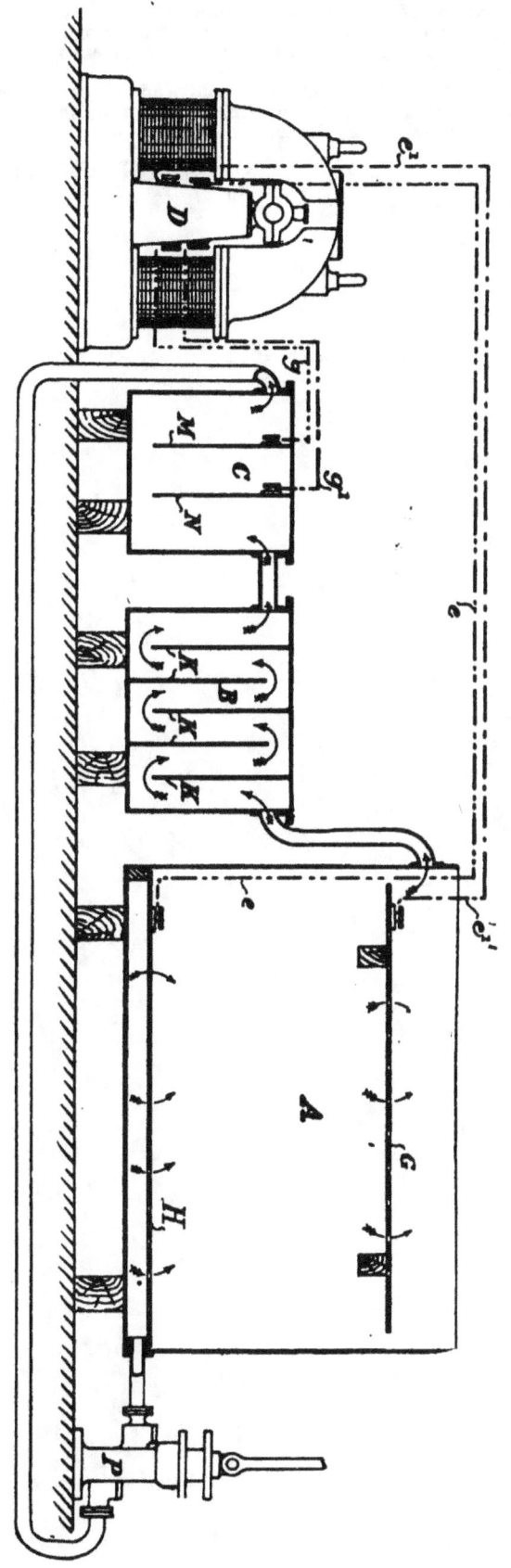

IMPROVED APPARATUS FOR THE EXTRACTION OF GOLD & SILVER FROM ORES.

—— C. M. Pelsticker's Patent. ——

6. In the above-described process of separating gold and silver from their ores in conjunction with an electric current, the use of a current of electricity of sufficient strength to decompose the double salt of cyanide of gold or silver and potassium without decomposing the cyanide of potassium itself.

7. In the above-described process of separating the gold and silver from their ores in conjunction with an electric current, the combination of an ore-tank with a settling-tank and a depositing-tank, substantially as described.

CARL PIELSTICKER.
By W. H. QUICK, His Agent.

Dated this 14th day of December, 1893.

UNITED STATES PATENT OFFICE.

WILLIAM DAVID JOHNSTON, of San Francisco, California.

METHOD OF ABSTRACTING GOLD AND SILVER FROM THEIR SOLUTIONS IN POTASSIUM CYANIDES.

Specification forming part of Letters Patent No. 522,260, dated July 3, 1894.

(Application filed November 20, 1893. Serial No. 491,473. No specimens.)

To all whom it may concern:

Be it known that I, William David Johnston, a citizen of the United States, residing in the City and County of San Francisco, State of California, have invented an Improvement in Methods of Abstracting Gold and Silver from their Solutions in Potassium Cyanide; and I hereby declare the following to be a full, clear, and exact description of the same:

Heretofore when solutions of gold and silver have been made in potassium cyanide, the metals have been recovered from their solution by the use of zinc in various forms.

The object of my invention is to recover the metals in a shorter time, and more economically, by the use of pulverized carbon, preferably in the form of charcoal.

To carry my invention into effect, I take carbon in a pulverized form as above, and place it upon suitable supports so as to form it into filters, through a series of which the cyanide liquid is caused to pass successively, leaving the metal deposited upon the carbon. The gold and silver are then recovered from the carbon by carefully burning the carbon, and smelting the residue with the usual fluxes. By thus employing a series of filters through which the solution is passed successively, I am able to recover upward of 95 per cent of the precious metal contained in the solution.

When only one filter is employed, only about one fourth of the gold can be extracted.

Having thus described my invention, what I claim as new, and desire to secure by Letters Patent, is—

1. The process of abstracting gold and silver from their solution in potassium cyanide, consisting in passing the liquid through a series of carbon filters within which the gold is arrested, substantially as described.

2. The process of abstracting gold and silver from their solution in potassium cyanide, consisting in passing the liquid through a series of carbon filters within which the gold is arrested, and then recovering the metal by burning the carbon and smelting the residue with suitable fluxes, substantially as described.

In witness whereof I have hereunto set my hand.

WILLIAM DAVID JOHNSTON.

Witnesses:
S. H. Nourse.
H. F. Ascheck.

Note.—The patents, where the country is not mentioned, are to be understood as being issued in Great Britain.

INDEX.

A

B

C

R

S

T

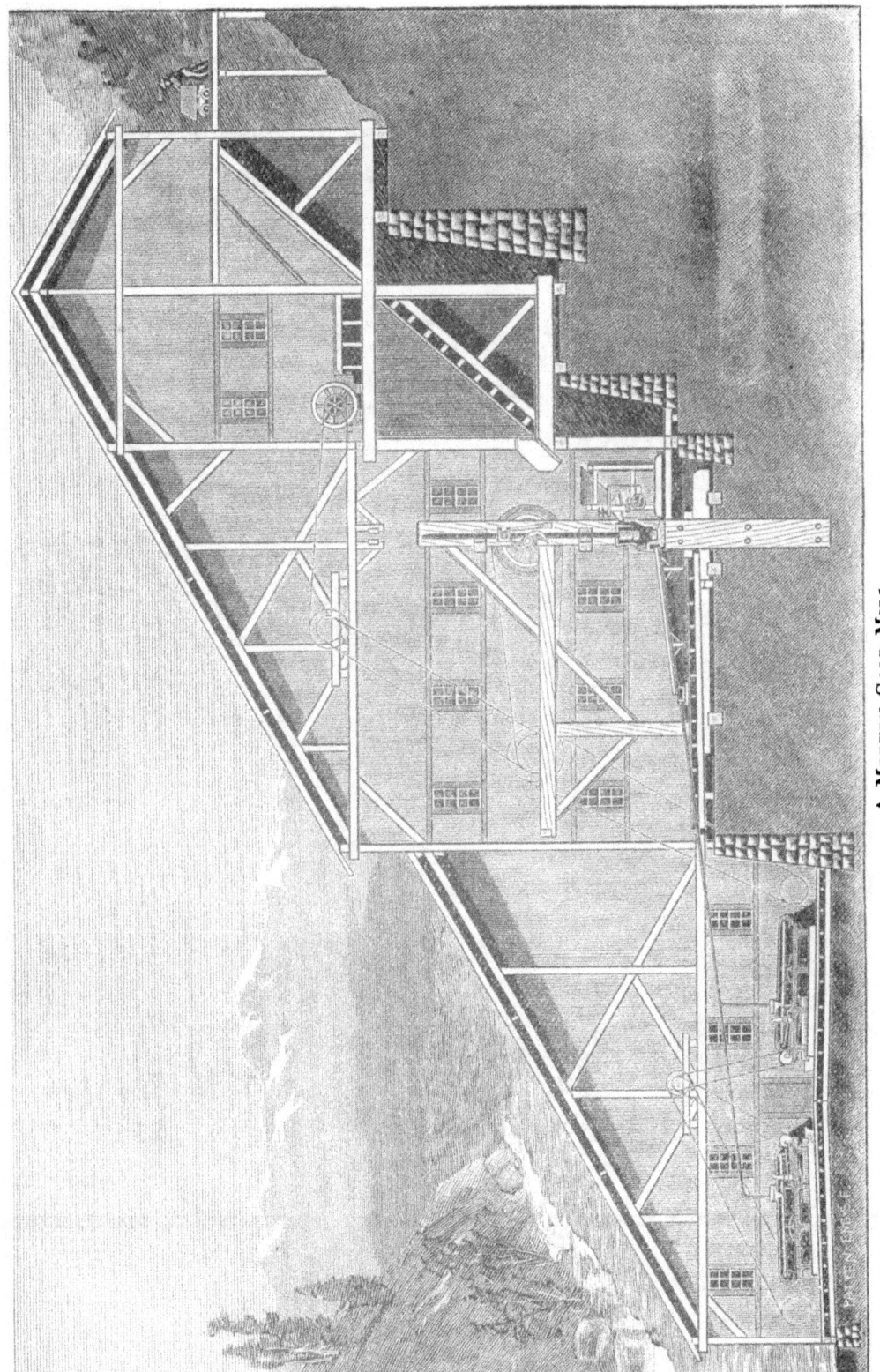

A Modern Gold Mill.

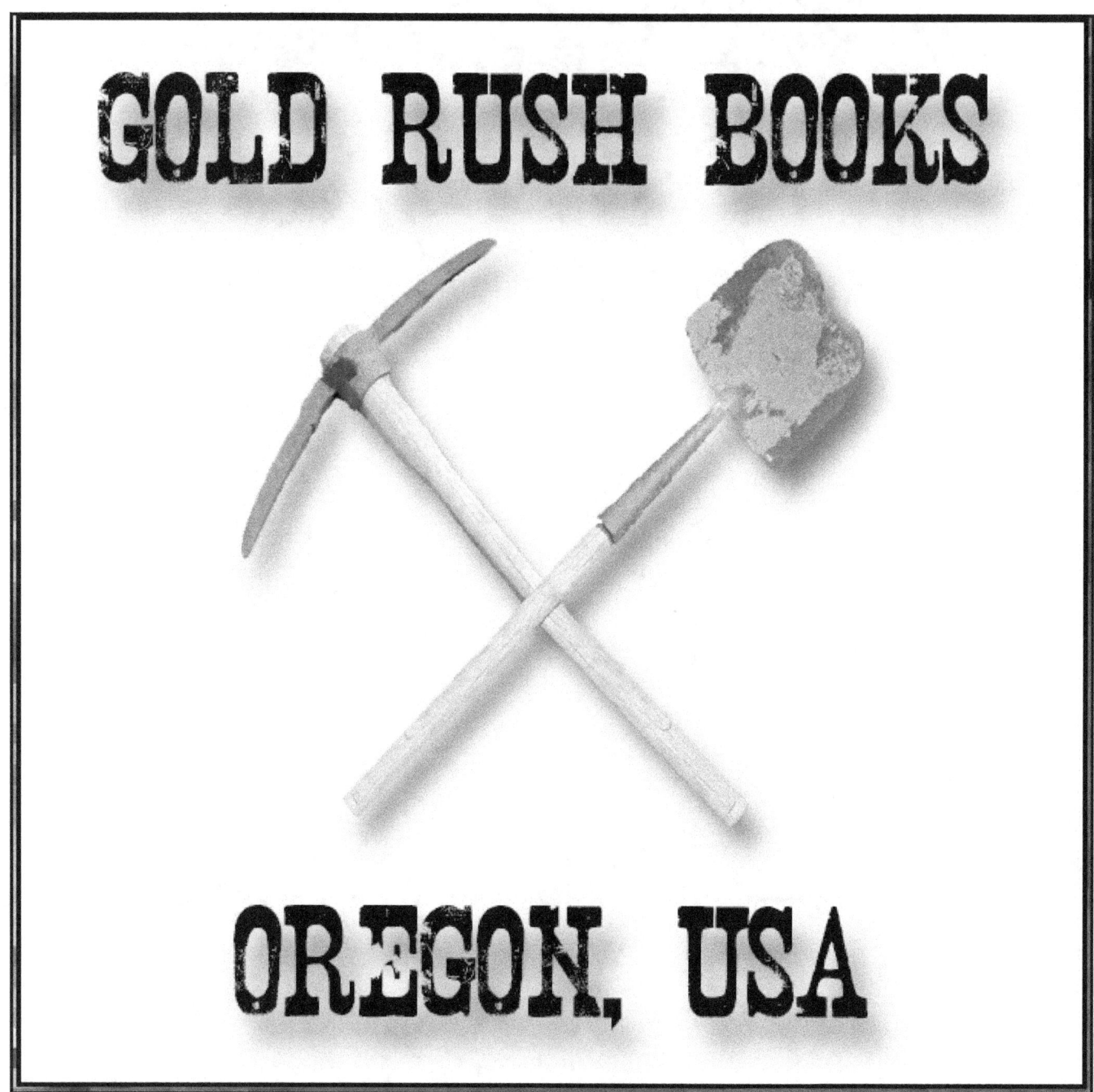

GOLD RUSH BOOKS

OREGON, USA

www.GoldMiningBooks.com

Books On Mining

Visit: www.goldminingbooks.com to order your copies or ask your favorite book seller to offer them.

Mining Books by Kerby Jackson

<u>Gold Dust: Stories From Oregon's Mining Years</u> - Oregon mining historian and prospector, Kerby Jackson, brings you a treasure trove of seventeen stories on Southern Oregon's rich history of gold prospecting, the prospectors and their discoveries, and the breathtaking areas they settled in and made homes. 5" X 8", 98 ppgs. Retail Price: $11.99

<u>The Golden Trail: More Stories From Oregon's Mining Years</u> - In his follow-up to "Gold Dust: Stories of Oregon's Mining Years", this time around, Jackson brings us twelve tales from Oregon's Gold Rush, including the story about the first gold strike on Canyon Creek in Grant County, about the old timers who found gold by the pail full at the Victor Mine near Galice, how Iradel Bray discovered a rich ledge of gold on the Coquille River during the height of the Rogue River War, a tale of two elderly miners on the hunt for a lost mine in the Cascade Mountains, details about the discovery of the famous Armstrong Nugget and others. 5" X 8", 70 ppgs. Retail Price: $10.99

Oregon Mining Books

<u>Geology and Mineral Resources of Josephine County, Oregon</u> - Unavailable since the 1970's, this important publication was originally compiled by the Oregon Department of Geology and Mineral Industries and includes important details on the economic geology and mineral resources of this important mining area in South Western Oregon. Included are notes on the history, geology and development of important mines, as well as insights into the mining of gold, copper, nickel, limestone, chromium and other minerals found in large quantities in Josephine County, Oregon.   8.5" X 11", 54 ppgs. Retail Price: $9.99

<u>Mines and Prospects of the Mount Reuben Mining District</u> - Unavailable since 1947, this important publication was originally compiled by geologist Elton Youngberg of the Oregon Department of Geology and Mineral Industries and includes detailed descriptions, histories and the geology of the Mount Reuben Mining District in Josephine County, Oregon. Included are notes on the history, geology, development and assay statistics, as well as underground maps of all the major mines and prospects in the vicinity of this much neglected mining district. 8.5" X 11", 48 ppgs. Retail Price: $9.99

<u>The Granite Mining District</u> - Notes on the history, geology and development of important mines in the well known Granite Mining District which is located in Grant County, Oregon. Some of the mines discussed include the Ajax, Blue Ribbon, Buffalo, Continental, Cougar-Independence, Magnolia, New York, Standard and the Tillicum. Also included are many rare maps pertaining to the mines in the area. 8.5" X 11", 48 ppgs. Retail Price: $9.99

<u>Ore Deposits of the Takilma and Waldo Mining Districts of Josephine County, Oregon</u> - The Waldo and Takilma mining districts are most notable for the fact that the earliest large scale mining of placer gold and copper in Oregon took place in these two areas. Included are details about some of the earliest large gold mines in the state such as the Llano de Oro, High Gravel, Cameron, Platerica, Deep Gravel and others, as well as copper mines such as the famous Queen of Bronze mine, the Waldo, Lily and Cowboy mines. This volume also includes six maps and 20 original illustrations. 8.5" X 11", 74 ppgs. Retail Price: $9.99

<u>Metal Mines of Douglas, Coos and Curry Counties, Oregon</u> - Oregon mining historian Kerby Jackson introduces us to a classic work on Oregon's mining history in this important re-issue of Bulletin 14C Volume 1, otherwise known as the Douglas, Coos & Curry Counties, Oregon Metal Mines Handbook. Unavailable since 1940, this important publication was originally compiled by the Oregon Department of Geology and Mineral Industries includes detailed descriptions, histories and the geology of over 250 metallic mineral mines and prospects in this rugged area of South West Oregon. 8.5" X 11", 158 ppgs. Retail Price: $19.99

Metal Mines of Jackson County, Oregon - Unavailable since 1943, this important publication was originally compiled by the Oregon Department of Geology and Mineral Industries includes detailed descriptions, histories and the geology of over 450 metallic mineral mines and prospects in Jackson County, Oregon. Included are such famous gold mining areas as Gold Hill, Jacksonville, Sterling and the Upper Applegate. **8.5" X 11"**, 220 ppgs. **Retail Price: $24.99**

Metal Mines of Josephine County, Oregon - Oregon mining historian Kerby Jackson introduces us to a classic work on Oregon's mining history in this important re-issue of Bulletin 14C, otherwise known as the Josephine County, Oregon Metal Mines Handbook. Unavailable since 1952, this important publication was originally compiled by the Oregon Department of Geology and Mineral Industries includes detailed descriptions, histories and the geology of over 500 metallic mineral mines and prospects in Josephine County, Oregon. **8.5" X 11"**, 250 ppgs. **Retail Price: $24.99**

Metal Mines of North East Oregon - Oregon mining historian Kerby Jackson introduces us to a classic work on Oregon's mining history in this important re-issue of Bulletin 14A and 14B, otherwise known as the North East Oregon Metal Mines Handbook. Unavailable since 1941, this important publication was originally compiled by the Oregon Department of Geology and Mineral Industries and includes detailed descriptions, histories and the geology of over 750 metallic mineral mines and prospects in North Eastern Oregon. **8.5" X 11"**, 310 ppgs. **Retail Price: $29.99**

Metal Mines of North West Oregon - Oregon mining historian Kerby Jackson introduces us to a classic work on Oregon's mining history in this important re-issue of Bulletin 14D, otherwise known as the North West Oregon Metal Mines Handbook. Unavailable since 1951, this important publication was originally compiled by the Oregon Department of Geology and Mineral Industries and includes detailed descriptions, histories and the geology of over 250 metallic mineral mines and prospects in North Western Oregon. **8.5" X 11"**, 182 ppgs. **Retail Price: $19.99**

Mines and Prospects of Oregon - Mining historian Kerby Jackson introduces us to a classic mining work by the Oregon Bureau of Mines in this important re-issue of The Handbook of Mines and Prospects of Oregon. Unavailable since 1916, this publication includes important insights into hundreds of gold, silver, copper, coal, limestone and other mines that operated in the State of Oregon around the turn of the 19th Century. Included are not only geological details on early mines throughout Oregon, but also insights into their history, production, locations and in some cases, also included are rare maps of their underground workings. **8.5" X 11"**, 314 ppgs. **Retail Price: $24.99**

Lode Gold of the Klamath Mountains of Northern California and South West Oregon
(See California Mining Books)

Mineral Resources of South West Oregon - Unavailable since 1914, this publication includes important insights into dozens of mines that once operated in South West Oregon, including the famous gold fields of Josephine and Jackson Counties, as well as the Coal Mines of Coos County. Included are not only geological details on early mines throughout South West Oregon, but also insights into their history, production and locations. **8.5" X 11"**, 154 ppgs. **Retail Price: $11.99**

Chromite Mining in The Klamath Mountains of California and Oregon
(See California Mining Books)

Southern Oregon Mineral Wealth - Unavailable since 1904, this rare publication provides a unique snapshot into the mines that were operating in the area at the time. Included are not only geological details on early mines throughout South West Oregon, but also insights into their history, production and locations. Some of the mining areas include Grave Creek, Greenback, Wolf Creek, Jump Off Joe Creek, Granite Hill, Galice, Mount Reuben, Gold Hill, Galls Creek, Kane Creek, Sardine Creek, Birdseye Creek, Evans Creek, Foots Creek, Jacksonville, Ashland, the Applegate River, Waldo, Kerby and the Illinois River, Althouse and Sucker Creek, as well as insights into local copper mining and other topics. **8.5" X 11"**, 64 ppgs. **Retail Price: $8.99**

Geology and Ore Deposits of the Takilma and Waldo Mining Districts - Unavailable since the 1933, this publication was originally compiled by the United States Geological Survey and includes details on gold and copper mining in the Takilma and Waldo Districts of Josephine County, Oregon. The Waldo and Takilma mining districts are most notable for the fact that the earliest large scale mining of placer gold and copper in Oregon took place in these two areas. Included in this report are details about some of the earliest large gold mines in the state such as the Llano de Oro, High Gravel, Cameron, Platerica, Deep Gravel and others, as well as copper mines such as the famous Queen of Bronze mine, the Waldo, Lily and Cowboy mines. In addition to geological examinations, insights are also provided into the production, day to day operations and early histories of these mines, as well as calculations of known mineral reserves in the area. This volume also includes six maps and 20 original illustrations. **8.5" X 11"**, 74 ppgs. **Retail Price: $9.99**

Gold Mines of Oregon - Oregon mining historian Kerby Jackson introduces us to a classic work on Oregon's mining history in this important re-issue of Bulletin 61, otherwise known as "Gold and Silver In Oregon". Unavailable since 1968, this important publication was originally compiled by geologists Howard C. Brooks and Len Ramp of the Oregon Department of Geology and Mineral Industries and includes detailed descriptions, histories and the geology of over 450 gold mines Oregon. Included are notes on the history, geology and gold production statistics of all the major mining areas in Oregon including the Klamath Mountains, the Blue Mountains and the North Cascades. While gold is where you find it, as every miner knows, the path to success is to prospect for gold where it was previously found. 8.5" X 11", 344 ppgs. **Retail Price: $24.99**

Mines and Mineral Resources of Curry County Oregon - Originally published in 1916, this important publication on Oregon Mining has not been available for nearly a century. Included are rare insights into the history, production and locations of dozens of gold mines in Curry County, Oregon, as well as detailed information on important Oregon mining districts in that area such as those at Agness, Bald Face Creek, Mule Creek, Boulder Creek, China Diggings, Collier Creek, Elk River, Gold Beach, Rock Creek, Sixes River and elsewhere. Particular attention is especially paid to the famous beach gold deposits of this portion of the Oregon Coast. 8.5" X 11", 140 ppgs. **Retail Price: $11.99**

Chromite Mining in South West Oregon - Originally published in 1961, this important publication on Oregon Mining has not been available for nearly a century. Included are rare insights into the history, production and locations of nearly 300 chromite mines in South Western Oregon. 8.5" X 11", 184 ppgs. **Retail Price: $14.99**

Mineral Resources of Douglas County Oregon - Originally published in 1972, this important publication on Oregon Mining has not been available for nearly forty years. Included are rare insights into the geology, history, production and locations of numerous gold mines and other mining properties in Douglas County, Oregon. 8.5" X 11", 124 ppgs. **Retail Price: $11.99**

Mineral Resources of Coos County Oregon - Originally published in 1972, this important publication on Oregon Mining has not been available for nearly forty years. Included are rare insights into the geology, history, production and locations of numerous gold mines and other mining properties in Coos County, Oregon. 8.5" X 11", 100 ppgs. **Retail Price: $11.99**

Mineral Resources of Lane County Oregon - Originally published in 1938, this important publication on Oregon Mining has not been available for nearly seventy five years. Included are extremely rare insights into the geology and mines of Lane County, Oregon, in particular in the Bohemia, Blue River, Oakridge, Black Butte and Winberry Mining Districts. 8.5" X 11", 82 ppgs. **Retail Price: $9.99**

Mineral Resources of the Upper Chetco River of Oregon: Including the Kalmiopsis Wilderness - Originally published in 1975, this important publication on Oregon Mining has not been available for nearly forty years. Withdrawn under the 1872 Mining Act since 1984, real insight into the minerals resources and mines of the Upper Chetco River has long been unavailable due to the remoteness of the area. Despite this, the decades of battle between property owners and environmental extremists over the last private mining inholding in the area has continued to pique the interest of those interested in mining and other forms of natural resource use. Gold mining began in the area in the 1850's and has a rich history in this geographic area, even if the facts surrounding it are little known. Included are twenty two rare photographs, as well as insights into the Becca and Morning Mine, the Emmly Mine (also known as Emily Camp), the Frazier Mine, the Golden Dream or Higgins Mine, Hustis Mine, Peck Mine and others. 8.5" X 11", 64 ppgs. **Retail Price: $8.99**

Gold Dredging in Oregon - Originally published in 1939, this important publication on Oregon Mining has not been available for nearly seventy five years. Included are extremely rare insights into the history and day to day operations of the dragline and bucketline gold dredges that once worked the placer gold fields of South West and North East Oregon in decades gone by. Also included are details into the areas that were worked by gold dredges in Josephine, Jackson, Baker and Grant counties, as well as the economic factors that impacted this mining method. This volume also offers a unique look into the values of river bottom land in relation to both farming and mining, in how farm lands were mined, re-soiled and reclamated after the dredges worked them. Featured are hard to find maps of the gold dredge fields, as well as rare photographs from a bygone era. 8.5" X 11", 86 ppgs. **Retail Price: $8.99**

Quick Silver Mining in Oregon - Originally published in 1963, this important publication on Oregon Mining has not been available for over fifty years. This publication includes details into the history and production of Elemental Mercury or Quicksilver in the State of Oregon. 8.5" X 11", 238 ppgs. **Retail Price: $15.99**

Mines of the Greenhorn Mining District of Grant County Oregon - Originally published in 1948, this important publication on Oregon Mining has not been available for over sixty five years. In this publication are rare insights into the mines of the famous Greenhorn Mining District of Grant County, Oregon, especially the famous Morning Mine. Also included are details on the Tempest, Tiger, Bi-Metallic, Windsor, Psyche, Big Johnny, Snow Creek, Banzette and Paramount Mines, as well as prospects in the vicinities in the famous mining areas of Mormon Basin, Vinegar Basin and Desolation Creek. Included are hard to find mine maps and dozens of rare photographs from the bygone era of Grant County's rich mining history. 8.5" X 11", 72 ppgs. **Retail Price: $9.99**

Geology of the Wallowa Mountains of Oregon: Part I (Volume 1) - Originally published in 1938, this important publication on Oregon Mining has not been available for nearly seventy five years. Included are details on the geology of this unique portion of North Eastern Oregon. This is the first part of a two book series on the area. Accompanying the text are rare photographs and historic maps. **8.5" X 11", 92 ppgs. Retail Price: $9.99**

Geology of the Wallowa Mountains of Oregon: Part II (Volume 2) - Originally published in 1938, this important publication on Oregon Mining has not been available for nearly seventy five years. Included are details on the geology of this unique portion of North Eastern Oregon. This is the first part of a two book series on the area. Accompanying the text are rare photographs and historic maps. **8.5" X 11", 94 ppgs. Retail Price: $9.99**

Field Identification of Minerals For Oregon Prospectors - Originally published in 1940, this important publication on Oregon Mining has not been available for nearly seventy five years. Included in this volume is an easy system for testing and identifying a wide range of minerals that might be found by prospectors, geologists and rockhounds in the State of Oregon, as well as in other locales. Topics include how to put together your own field testing kit and how to conduct rudimentary tests in the field. This volume is written in a clear and concise way to make it useful even for beginners. **8.5" X 11", 158 ppgs. Retail Price: $14.99**

The Bohemia Mining District of Oregon - Originally published in 1900, this important publication on Oregon Mining has not been available for over a century. Included in this volume are important insights into the famous Bohemia Mining District of Oregon, including the histories and locations of important gold mines in the area such as the Ophir Mine, Clarence, Acturas, Peek-a-boo, White Swan, Combination Mine, the Musick Mine, The California, White Ghost, The Mystery, Wall Street, Vesuvius, Story, Lizzie Bullock, Delta, Elsie Dora, Golden Slipper, Broadway, Champion Mine, Knott, Noonday, Helena, White Wings, Riverside and others. Also included are notes on the nearby Blue River Mining District. **8.5" X 11", 58 ppgs. Retail Price: $9.99**

The Gold Fields of Eastern Oregon - Unavailable since 1900, this publication was originally compiled by the Baker City Chamber of Commerce Offering important insights into the gold mining history of Eastern Oregon, "The Gold Fields of Eastern Oregon" sheds a rare light on many of the gold mines that were operating at the turn of the 19th Century in Baker County and Grant County in North Eastern Oregon. Some of the areas featured include the Cable Cove District, Baisely-Elhorn, Granite, Red Boy, Bonanza, Susanville, Sparta, Virtue, Vaughn, Sumpter, Burnt River, Rye Valley and other mining districts. Included is basic information on not only many gold mines that are well known to those interested in Eastern Oregon mining history, but also many mines and prospects which have been mostly lost to the passage of time. Accompanying are numerous rare photos **8.5" X 11", 78 ppgs. Retail Price: $10.99**

Gold Mining in Eastern Oregon - Originally published in 1938, this important publication on Oregon Mining has not been available for over a century. Included in this volume are important insights into the famous mining districts of Eastern Oregon during the late 1930's. Particular attention is given to those gold mines with milling and concentrating facilities in the Greenhorn, Red Boy, Alamo, Bonanza, Granite, Cable Cove, Cracker Creek, Virtue, Keating, Medical Springs, Sanger, Sparta, Chicken Creek, Mormon Basin, Connor Creek, Cornucopia and the Bull Run Mining Districts. Some of the mines featured include the Ben Harrison, North Pole-Columbia, Highland Maxwell, Baisley-Elkhorn, White Swan, Balm Creek, Twin Baby, Gem of Sparta, New Deal, Gleason, Gifford-Johnson, Cornucopia, Record, Bull Run, Orion and others. Of particular interest are the mill flow sheets and descriptions of milling operations of these mines. **8.5" X 11", 68 ppgs. Retail Price: $8.99**

The Gold Belt of the Blue Mountains of Oregon - Originally published in 1901, this important publication on Oregon Mining has not been available for over a century. Included in this volume are rare insights into the gold deposits of the Blue Mountains of North East Oregon, including the history of their early discovery and early production. Extensive details are offered on this important mining area's mineralogy and economic geology, as well as insights into nearby gold placers, silver deposits and copper deposits. Featured are the Elkhorn and Rock Creek mining districts, the Pocahontas district, Auburn and Minersville districts, Sumpter and Cracker Creek, Cable Cove, the Camp Carson district, Granite, Alamo, Greenhorn, Robinsonville, the Upper Burnt River Valley and Bonanza districts, Susanville, Quartzburg, Canyon Creek, Virtue, the Copper Butte district, the North Powder River, Sparta, Eagle Creek, Cornucopia, Pine Creek, Lower Powder River, the Upper Snake River Canyon, Rye Valley, Lower Burnt River Valley, Mormon Basin, the Malheur and Clarks Creek districts, Sutton Creek and others. Of particular interest are important details on numerous gold mines and prospects in these mining districts, including their locations, histories, geology and other important information, as well as information on silver, copper and fire opal deposits. **8.5" X 11", 250 ppgs. Retail Price: $24.99**

<u>Mining in the Cascades Range of Oregon</u> - Originally published in 1938, this important publication on Oregon Mining has not been available for over seventy five years. Included in this volume are rare insights into the gold mines and other types of metal mines in the Cascades Mountain Range of Oregon. Some of the important mining areas covered include the famous Bohemia Mining District, the North Santiam Mining District, Quartzville Mining District, Blue River Mining District, Fall Creek Mining District, Oakridge District, Zinc District, Buzzard-Al Sarena District, Grand Cove, Climax District and Barron Mining District. Of particular interest are important details on over 100 mines and prospects in these mining districts, including their locations, histories, geology and other important information. **8.5" X 11", 170 ppgs. Retail Price: $14.99**

Idaho Mining Books

<u>Gold in Idaho</u> - Unavailable since the 1940's, this publication was originally compiled by the Idaho Bureau of Mines and includes details on gold mining in Idaho. Included is not only raw data on gold production in Idaho, but also valuable insight into where gold may be found in Idaho, as well as practical information on the gold bearing rocks and other geological features that will assist those looking for placer and lode gold in the State of Idaho. This volume also includes thirteen gold maps that greatly enhance the practical usability of the information contained in this small book detailing where to find gold in Idaho. **8.5" X 11", 72 ppgs. Retail Price: $9.99**

<u>Geology of the Couer D'Alene Mining District of Idaho</u> - Unavailable since 1961, this publication was originally compiled by the Idaho Bureau of Mines and Geology and includes details on the mining of gold, silver and other minerals in the famous Coeur D'Alene Mining District in Northern Idaho. Included are details on the early history of the Coeur D'Alene Mining District, local tectonic settings, ore deposit features, information on the mineral belts of the Osburn Fault, as well as detailed information on the famous Bunker Hill Mine, the Dayrock Mine, Galena Mine, Lucky Friday Mine and the infamous Sunshine Mine. This volume also includes sixteen hard to find maps. **8.5" X 11", 70 ppgs. Retail Price: $9.99**

<u>The Gold Camps and Silver Cities of Idaho</u> - Originally published in 1963, this important publication on Idaho Mining has not been available for nearly fifty years. Included are rare insights into the history of Idaho's Gold Rush, as well as the mad craze for silver in the Idaho Panhandle. Documented in fine detail are the early mining excitements at Boise Basin, at South Boise, in the Owyhees, at Deadwood, Long Valley, Stanley Basin and Robinson Bar, at Atlanta, on the famous Boise River, Volcano, Little Smokey, Banner, Boise Ridge, Hailey, Leesburg, Lemhi, Pearl, at South Mountain, Shoup and Ulysses, Yellow Jacket and Loon Creek. The story follows with the appearance of Chinese miners at the new mining camps on the Snake River, Black Pine, Yankee Fork, Bay Horse, Clayton, Heath, Seven Devils, Gibbonsville, Vienna and Sawtooth City. Also included are special sections on the Idaho Lead and Silver mines of the late 1800's, as well as the mining discoveries of the early 1900's that paved the way for Idaho's modern mining and mineral industry. Lavishly illustrated with rare historic photos, this volume provides a one of a kind documentary into Idaho's mining history that is sure to be enjoyed by not only modern miners and prospectors who still scour the hills in search of nature's treasures, but also those enjoy history and tromping through overgrown ghost towns and long abandoned mining camps. **8.5" X 11", 186 ppgs. Retail Price: $14.99**

<u>Ore Deposits and Mining in North Western Custer County Idaho</u> - Unavailable since 1913, this important publication was originally published by the Us Department of the Interior and has been unavailable for a century. Included are fine details on the geology, geography, gold placers and gold and silver bearing quartz veins of the mining region of North West Custer County, Idaho. Of particular interest is a rare look at the mines and prospects of the region, including those such as the Ramshorn Mine, SkyLark, Riverview, Excelsior, Beardsley, Pacific, Hoosier, Silver Brick, Forest Rose and dozens of others in the Bay Horse Mining District. Also covered are the mines of the Yankee Fork District such as the Lucky Boy, Badger, Black, Enterprise, Charles Dickens, Morrison, Golden Sunbeam, Montana, Golden Gate and others, as well as those in the Loon Mining District. **8.5" X 11", 126 ppgs. Retail Price: $12.99**

<u>Gold Rush To Idaho</u> - Unavailable since 1963, this important publication was originally published by the Idaho Bureau of Mines and has been unavailable for 50 years. "Gold Rush To Idaho" revisits the earliest years of the discovery of gold in Idaho Territory and introduces us to the conditions that the pioneer gold seekers met when they blazed a trail through the wilderness of Idaho's mountains and discovered the precious yellow metal at Oro Fino and Pierce. Subsequent rushes followed at places like Elk City, Newsome, Clearwater Station, Florence, Warrens and elsewhere. Of particular interest is a rare look at the hardships that the first miners in Idaho met with during their day to day existences and their attempts to bring law and order to their mining camps. **8.5" X 11", 88 ppgs. Retail Price: $9.99**

The Geology and Mines of Northern Idaho and North Western Montana - Unavailable since 1909, this important publication was originally published by the Us Department of the Interior and has been unavailable for a century. Included are fine details on the geology and geography of the mining regions of Northern Idaho and North Western Montana. Of particular interest is a rare look at the mines and prospects of the region, including those in the Pine Creek Mining District, Lake Pend Oreille district, Troy Mining District, Sylvanite District, Cabinet Mining District, Prospect Mining District and the Missoula Valley. Some of the mines featured include the Iron Mountain, Silver Butte, Snowshoe, Grouse Mountain Mine and others. **8.5″ X 11″, 142 ppgs. Retail Price: $12.99**

Utah Mining Books

Fluorite in Utah - Unavailable since 1954, this publication was originally compiled by the USGS, State of Utah and U.S. Atomic Energy Commission and details the mining of fluorspar, also known as fluorite in the State of Utah. Included are details on the geology and history of fluorspar (fluorite) mining in Utah, including details on where this unique gem mineral may be found in the State of Utah. **8.5″ X 11″, 60 ppgs. Retail Price: $8.99**

California Mining Books

The Tertiary Gravels of the Sierra Nevada of California - Mining historian Kerby Jackson introduces us to a classic mining work by Waldemar Lindgren in this important re-issue of The Tertiary Gravels of the Sierra Nevada of California. Unavailable since 1911, this publication includes details on the gold bearing ancient river channels of the famous Sierra Nevada region of California. **8.5″ X 11″, 282 ppgs. Retail Price: $19.99**

The Mother Lode Mining Region of California - Unavailable since 1900, this publication includes details on the gold mines of California's famous Mother Lode gold mining area. Included are details on the geology, history and important gold mines of the region, as well as insights into historic mining methods, mine timbering, mining machinery, mining bell signals and other details on how these mines operated. Also included are insights into the gold mines of the California Mother Lode that were in operation during the first sixty years of California's mining history. **8.5″ X 11″, 176 ppgs. Retail Price: $14.99**

Lode Gold of the Klamath Mountains of Northern California and South West Oregon - Unavailable since 1971, this publication was originally compiled by Preston E. Hotz and includes details on the lode mining districts of Oregon and California's Klamath Mountains. Included are details on the geology, history and important lode mines of the French Gulch, Deadwood, Whiskeytown, Shasta, Redding, Muletown, South Fork, Old Diggings, Dog Creek (Delta), Bully Choop (Indian Creek), Harrison Gulch, Hayfork, Minersville, Trinity Center, Canyon Creek, East Fork, New River, Denny, Liberty (Black Bear), Cecilville, Callahan, Yreka, Fort Jones and Happy Camp mining districts in California, as well as the Ashland, Rogue River, Applegate, Illinois River, Takilma, Greenback, Galice, Silver Peak, Myrtle Creek and Mule Creek districts of South Western Oregon. Also included are insights into the mineralization and other characteristics of this important mining region. **8.5″ X 11″, 100 ppgs. Retail Price: $10.99**

Mines and Mineral Resources of Shasta County, Siskiyou County, Trinity County: California - Unavailable since 1915, this publication was originally compiled by the California State Mining Bureau and includes details on the gold mines of this area of Northern California. Also included are insights into the mineralization and other characteristics of this important mining region, as well as the location of historic gold mines. **8.5″ X 11″, 204 ppgs. Retail Price: $19.99**

Geology of the Yreka Quadrangle, Siskiyou County, California - Unavailable since 1977, this publication was originally compiled by Preston E. Hotz and includes details on the geology of the Yreka Quadrangle of Siskiyou County, California. Also included are insights into the mineralization and other characteristics of this important mining region. **8.5″ X 11″, 78 ppgs. Retail Price: $7.99**

Mines of San Diego and Imperial Counties, California - Originally published in 1914, this important publication on California Mining has not been available for a century. This publication includes important information on the early gold mines of San Diego and Imperial County, which were some of the first gold fields mined in California by early Spanish and Mexican miners before the 49ers came on the scene. Included are not only details on early mining methods in the area, production statistics and geological information, but also the location of the early gold mines that helped make California "The Golden State". Also included are details on the mining of other minerals such as silver, lead, zinc, manganese, tungsten, vanadium, asbestos, barite, borax, cement, clay, dolomite, fluospar, gem stones, graphite, marble, salines, petroleum, strontium, talc and others. **8.5″ X 11″, 116 ppgs. Retail Price: $12.99**

Mines of Sierra County, California - Unavailable since 1920, this publication was originally compiled by the California State Mining Bureau and includes details on the gold mines of Sierra County, California. Also included are insights into the mineralization and other characteristics of this important mining region, as well as the location of historic gold mines. **8.5″ X 11″, 156 ppgs. Retail Price: $19.99**

Mines of Plumas County, California - Unavailable since 1918, this publication was originally compiled by the California State Mining Bureau and includes details on the gold mines of Plumas County, California. Also included are insights into the mineralization and other characteristics of this important mining region, as well as the location of historic gold mines. 8.5" X 11", 200 ppgs. Retail Price: $19.99

Mines of El Dorado, Placer, Sacramento and Yuba Counties, California - Originally published in 1917, this important publication on California Mining has not been available for nearly a century. This publication includes important information on the early gold mines of El Dorado County, Placer County, Sacramento County and Yuba County, which were some of the first gold fields mined by the Forty-Niners during the California Gold Rush. Included are not only details on early mining methods in the area, production statistics and geological information, but also the location of the early gold mines that helped make California "The Golden State". Also included are insights into the early mining of chrome, copper and other minerals in this important mining area. 8.5" X 11", 204 ppgs. Retail Price: $19.99

Mines of Los Angeles, Orange and Riverside Counties, California - Originally published in 1917, this important publication on California Mining has not been available for nearly a century. This publication includes important information on the early gold mines of Los Angeles County, Orange County and Riverside County, which were some of the first gold fields mined in California by early Spanish and Mexican miners before the 49ers came on the scene. Included are not only details on early mining methods in the area, production statistics and geological information, but also the location of the early gold mines that helped make California "The Golden State". 8.5" X 11", 146 ppgs. Retail Price: $12.99

Mines of San Bernadino and Tulare Counties, California - Originally published in 1917, this important publication on California Mining has not been available for nearly a century. This publication includes important information on the early gold mines of San Bernadino and Tulare County, which were some of the first gold fields mined in California by early Spanish and Mexican miners before the 49ers came on the scene. Included are not only details on early mining methods in the area, production statistics and geological information, but also the location of the early gold mines that helped make California "The Golden State". Also included are details on the mining of other minerals such as copper, iron, lead, zinc, manganese, tungsten, vanadium, asbestos, barite, borax, cement, clay, dolomite, fluospar, gem stones, graphite, marble, salines, petroleum, stronium, talc and others. 8.5" X 11", 200 ppgs. Retail Price: $19.99

Chromite Mining in The Klamath Mountains of California and Oregon - Unavailable since 1919, this publication was originally compiled by J.S. Diller of the United States Department of Geological Survey and includes details on the chromite mines of this area of Northern California and Southern Oregon. Also included are insights into the mineralization and other characteristics of this important mining region, as well as the location of historic mines. Also included are insights into chromite mining in Eastern Oregon and Montana. 8.5" X 11", 98 ppgs. Retail Price: $9.99

Mines and Mining in Amador, Calaveras and Tuolumne Counties, California - Unavailable since 1915, this publication was originally compiled by William Tucker and includes details on the mines and mineral resources of this important California mining area. Included are details on the geology, history and important gold mines of the region, as well as insights into other local mineral resources such as asbestos, clay, copper, talc, limestone and others. Also included are insights into the mineralization and other characteristics of this important portion of California's Mother Lode mining region. 8.5" X 11", 198 ppgs. Retail Price: $14.99

The Cerro Gordo Mining District of Inyo County California - Unavailable since 1963, this publication was originally compiled by the United States Department of Interior. Included are insights into the mineralization and other characteristics of this important mining region of Southern California. Topics include the mining of gold and silver in this important mining district in Inyo County, California, including details on the history, production and locations of the Cerro Gordo Mine, the Morning Star Mine, Estelle Tunnel, Charles Lease Tunnel, Ignacio, Hart, Crosscut Tunnel, Sunset, Upper Newtown, Newtown, Ella, Perseverance, Newsboy, Belmont and other silver and gold mines in the Cerro Gordo Mining District. This volume also includes important insights into the fossil record, geologic formations, faults and other aspects of economic geology in this California mining district. 8.5" X 11", 104 ppgs. Retail Price: $10.99

Mining in Butte, Lassen, Modoc, Sutter and Tehama Counties of California - Unavailable since 1917, this publication was originally compiled by the United States Department of Interior. Included are insights into the mineralization and other characteristics of this important mining region of California. Topics include the mining of asbestos, chromite, gold, diamonds and manganese in Butte County, the mining of gold and copper in the Hayden Hill and Diamond Mountain mining districts of Lassen County, the mining of coal, salt, copper and gold in the High Grade and Winters mining districts of Modoc County, gold mining in Sutter County and the mining of gold, chromite, manganese and copper in Tehama County. This volume also includes the production records and locations of numerous mines in this important mining region. 8.5" X 11", 114 ppgs. Retail Price: $11.99

Mines of Trinity County California - Originally published in 1965, this important publication on California Mining has not been available for nearly fifty years. This publication includes important information on mines and mining in Trinity County, California, as well insights into the mineralization and geology of this important mining area in Northern California. Included are extensive details on hardrock and placer gold mines and prospects, including charts showing the locations of these historic mines.. 8.5" X 11", 144 ppgs. Retail Price: $12.99

Mines of Kern County California - Originally published in 1962, this important publication on California Mining has not been available for nearly fifty years. This publication includes important information on mines and mining in Kern County, California, as well insights into the mineralization and geology of this important mining area in California. Included are extensive details on hardrock and placer gold mines and prospects, including charts showing the locations of these historic mines. 8.5" X 11", 398 ppgs. Retail Price: $24.99

Mines of Calaveras County California - Originally published in 1962, this important publication on California Mining has not been available for nearly fifty years. This publication includes important information on mines and mining in Calaveras County, California, as well insights into the mineralization and geology of this important mining area in Northern California. Included are extensive details on hardrock and placer gold mines and prospects, including charts showing the locations of these historic mines. 8.5" X 11", 236 ppgs. Retail Price: $19.99

Lode Gold Mining in Grass Valley California - Unavailable since 1940, this publication was originally compiled by the United States Department of Interior. Included are insights into the gold mineralization and other characteristics of this important mining region of Nevada County, California. This volume also includes important insights into the geologic formations, faults and other aspects of economic geology in this California mining district. Of particular interest are the fine details on many hardrock gold mines in the area, including their locations, histories, development and mineralization. Some of the mines featured include the Gold Hill Mine, Massachusetts Hill, Boundary, Peabody, Golden Center, North Star, Omaha, Lone Jack, Homeward Bound, Hartery, Wisconsin, Allison Ranch, Phoenix, Kate Hayes, W.Y.O.D., Empire, Rich Hill, Daisy Hill, Orleans, Sultana, Centennial, Conlin, Ben Franklin, Crown Point and many others. 8.5" X 11", 148 ppgs. Retail Price: $12.99

Alaska Mining Books

Ore Deposits of the Willow Creek Mining District, Alaska - Unavailable since 1954, this hard to find publication includes valuable insights into the Willow Creek Mining District near Hatcher Pass in Alaska. The publication includes insights into the history, geology and locations of the well known mines in the area, including the Gold Cord, Independence, Fern, Mabel, Lonesome, Snowbird, Schroff-O'Neil, High Grade, Marion Twin, Thorpe, Webfoot, Kelly-Willow, Lane, Holland and others. 8.5" X 11", 96 ppgs. Retail Price: $9.99

Arizona Mining Books

Mines and Mining in Northern Yuma County Arizona - Originally published in 1911, this important publication on Arizona Mining has not been available for over a hundred years. Included are rare insights into the gold, silver, copper and quicksilver mines of Yuma County, Arizona together with hard to find maps and photographs. Some of the mines and mining districts featured include the Planet Copper Mine, Mineral Hill, the Clara Consolidated Mine, Viati Mine, Copper Basin prospect, Bowman Mine, Quartz King, Billy Mack, Carnation, the Wardwell and Osbourne, Valensuella Copper, the Mariquita, Colonial Mine, the French American, the New York-Plomosa, Guadalupe, Lead Camp, Mudersbach Copper Camp, Yellow Bird, the Arizona Northern (Salome Strike), Bonanza (Harqua Hala), Golden Eagle, Hercules, Socorro and others. 8.5" X 11", 144 ppgs. Retail Price: $11.99

The Aravaipa and Stanley Mining Districts of Graham County Arizona - Originally published in 1925, this important publication on Arizona Mining has not been available for nearly ninety years. Included are rare insights into the gold and silver mines of these two important mining districts, together with hard to find maps. 8.5" X 11", 140 ppgs. Retail Price: $11.99

Gold in the Gold Basin and Lost Basin Mining Districts of Mohave County, Arizona - This volume contains rare insights into the geology and gold mineralization of the Gold Basin and Lost Basin Mining Districts of Mohave County, Arizona that will be of benefit to miners and prospectors. Also included is a significant body of information on the gold mines and prospects of this portion of Arizona. This volume is lavishly illustrated with rare photos and mining maps. 8.5" X 11", 188 ppgs. Retail Price: $19.99

Mines of the Jerome and Bradshaw Mountains of Arizona - This important publication on Arizona Mining has not been available for ninety years. This volume contains rare insights into the geology and ore deposits of the Jerome and Bradshaw Mountains of Arizona that will be of benefit to miners and prospectors who work those areas. Included is a significant body of information on the mines and prospects of the Verde, Black Hills, Cherry Creek, Prescott, Walker, Groom Creek, Hassayampa, Bigbug, Turkey Creek, Agua Fria, Black Canyon, Peck, Tiger, Pine Grove, Bradshaw, Tintop, Humbug and Castle Creek Mining Districts. This volume is lavishly illustrated with rare photos and mining maps. 8.5" X 11", 218 ppgs. Retail Price: $19.99

The Ajo Mining District of Pima County Arizona - This important publication on Arizona Mining has not been available for nearly seventy years. This volume contains rare insights into the geology and mineralization of the Ajo Mining District in Pima County, Arizona and in particular the famous New Cornelia Mine. 8.5″ X 11″, 126 ppgs. Retail Price: $11.99

Mining in the Santa Rita and Patagonia Mountains of Arizona - Originally published in 1915, this important publication on Arizona Mining has not been available for nearly a century. Included are rare insights into hundreds of gold, silver, copper and other mines in this famous Arizona mining area. Details include the locations, geology, history, production and other facts of the mines of this region. 8.5″ X 11″, 394 ppgs. Retail Price: $24.99

Montana Mining Books

A History of Butte Montana: The World's Greatest Mining Camp - First published in 1900 by H.C. Freeman, this important publication sheds a bright light on one of the most important mining areas in the history of The West. Together with his insights, as well as rare photographs of the periods, Harry Freeman describes Butte and its vicinity from its early beginnings, right up to its flush years when copper flowed from its mines like a river. At the time of publication, Butte, Montana was known worldwide as "The Richest Mining Spot On Earth" and produced not only vast amounts of copper, but also silver, gold and other metals from its mines. Freeman illustrates, with great detail, the most important mines in the vicinity of Butte, providing rare details on their owners, their history and most importantly, how the mines operated and how their treasures were extracted. Of particular interest are the dozens of rare photographs that depict mines such as the famous Anaconda, the Silver Bow, the Smoke House, Moose, Paulin, Buffalo, Little Minah, the Mountain Consolidated, West Greyrock, Cora, the Green Mountain, Diamond, Bell, Parnell, the Neversweat, Nipper, Original and many others. 8.5″ X 11″, 142 ppgs. Retail Price: $12.99

The Butte Mining District of Montana - This important publication on Montana Mining has not been available for over a century. Included are rare insights into the gold, copper and silver mines of Butte, Montana together with hard to find maps and photographs. Some of the topics include the early history of gold, silver and copper mining in the Butte area, insight into the geology of its mining areas, the local distribution of gold, silver and copper ores, as well their composition and how to identify them. Also included are detailed facts about the mines in the Butte Mining District, including the famous Anaconda Mine, Gagnon, Parrot, Blue Vein, Moscow, Poulin, Stella, Buffalo, Green Mountain, Wake Up Jim, the Diamond-Bell Group, Mountain Consolidated, East Greyrock, West Greyrock, Snowball, Corra, Speculator, Adirondack, Miners Union, the Jessie-Edith May Group, Otisco, Iduna, Colorado, Lizzie, Cambers, Anderson, Hesperus, Preferencia and dozens of others. 8.5″ X 11″, 298 ppgs. Retail Price: $24.99

Mines of the Helena Mining Region of Montana - This important publication on Montana Mining has not been available for over a century. Included are rare insights into the gold, copper and silver mines of the vicinity of Helena, Montana, including the Marysville Mining District, Elliston Mining District, Rimini Mining District, Helena Mining District, Clancy Mining District, Wickes Mining District, Boulder and Basin Mining Districts and the Elkhorn Mining District. Some of the topics include the early history of gold, silver and copper mining in the Helena area, insight into the geology of its mining areas, the local distribution of gold, silver and copper ores, as well their composition and how to identify them. Also included are detailed facts, history, geology and locations of over one hundred gold, silver and copper mines in the area . 8.5″ X 11″, 162 ppgs, Retail Price: $14.99

Mines and Geology of the Garnet Range of Montana - This important publication on Montana Mining has not been available for over a century. Included are rare insights into the gold, copper and silver mines of the vicinity of this important mining area of Montana. Some of the topics include the early history of gold, silver and copper mining in the Garnet Mountains, insight into the geology of its mining areas, the local distribution of gold, silver and copper ores, as well their composition and how to identify them. Also included are detailed facts, history, geology and locations of numerous gold, silver and copper mines in the area . 8.5″ X 11″, 100 ppgs, Retail Price: $11.99

Mines and Geology of the Philipsburg Quadrangle of Montana - This important publication on Montana Mining has not been available for over a century. Included are rare insights into the gold, copper and silver mines of the vicinity of this important mining area of Montana. Some of the topics include the early history of gold, silver and copper mining in the Philipsburg Quadrangle, insight into the geology of its mining areas, the local distribution of gold, silver and copper ores, as well their composition and how to identify them. Also included are detailed facts, history, geology and locations of over one hundred gold, silver and copper mines in the area 8.5″ X 11″, 290 ppgs, Retail Price: $24.99

Geology of the Marysville Mining District of Montana - Included are rare insights into the mining geology of the Marysville Mining District. Some of the topics include the early history of gold, silver and copper mining in the area, insight into the geology of its mining areas, the local distribution of gold, silver and copper ores, as well their composition and how to identify them. Also included are detailed facts, history, geology and locations of gold, silver and copper mines in the area 8.5″ X 11″, 198 ppgs, Retail Price: $19.99

The Geology and Mines of Northern Idaho and North Western Montana

See listing under Idaho.

Nevada Mining Books

The Bull Frog Mining District of Nevada - Unavailable since 1910, this publication was originally compiled by the United States Department of Interior. This volume also includes important insights into the geologic formations, faults and other aspects of economic geology in this Nevada mining district. Of particular interest are the fine details on many mines in the area, including their locations, histories, development and mineralization. Some of the mines featured include the National Bank Mine, Providence, Gibraltor, Tramps, Denver, Original Bullfrog, Gold Bar, Mayflower, Homestake-King and other mines and prospects. **8.5″ X 11″, 152 ppgs, Retail Price: $14.99**

Colorado Mining Books

Ores of The Leadville Mining District - Unavailable since 1926, this publication was originally compiled by the United States Department of Interior. This volume also includes important insights into the ores and mineralization of the Leadville Mining District in Colorado. Topics include historic ore prospecting methods, local geology, insights into ore veins and stockworks, the local trend and distribution of ore channels, reverse faults, shattered rock above replacement ore bodies, mineral enrichment in oxidized and sulphide zones and more. **8.5″ X 11″, 66 ppgs, Retail Price: $8.99**

Mining in Colorado - Unavailable since 1926, this publication was originally compiled by the United States Department of Interior. This volume also includes important insights into the mining history of Colorado from its early beginnings in the 1850's right up to the mid 1920's. Not only is Colorado's gold mining heritage included, but also its silver, copper, lead and zinc mining industry. Each mining area is treated separately, detailing the development of Colorado's mines on a county by county basis. **8.5″ X 11″, 284 ppgs, Retail Price: $19.99**

Gold Mining in Gilpin County Colorado - Unavailable since 1876, this publication was originally compiled by the Register Steam Printing House of Central City, Colorado. A rare glimpse at the gold mining history and early mines of Gilpin County, Colorado from their first discovery in the 1850's up to the "flush years" of the mid 1870's. Of particular interest is the history of the discovery of gold in Gilpin County and details about the men who made those first strikes. Special focus is given to the early gold mines and first mining districts of the area, many of which are not detailed in other books on Colorado's gold mining history. **8.5″ X 11″, 156 ppgs, Retail Price: $12.99**

Mining in the Gold Brick Mining District of Colorado - Important insights into the history of the Gold Brick Mining District, as well as its local geography and economic geology. Also included are the histories and locations of historic mines in this important Colorado Mining District, including the Cortland, Carter, Raymond, Gold Links, Sacramento, Bassick, Sandy Hook, Chronicle, Grand Prize, Chloride, Granite Mountain, Lucille, Gray Mountain, Hilltop, Maggie Mitchell, Silver Islet, Revenue, Roosevelt, Carbonate King and others. In addition to hardrock mining, are also included are details on gold placer mining in this portion of Colorado. **8.5″ X 11″, 140 ppgs, Retail Price: $12.99**

Washington Mining Books

The Republic Mining District of Washington - Unavailable since 1910, this important publication was originally published by the Washington Geologic Survey and has been unavailable for a century. Topics include the geology, rock formations and the formation of ore deposits in this important mining area of Washington State. Also included are hard to find details on the geology, history and locations of dozens of mines in the area. Some of the mines featured include the New Republic Mine, Ben Hur, Morning Glory, the South Republic Mine, Quilp, Surprise, Black Tail, Lone Pine, San Poil, Mountain Lion, Tom Thumb, Elcaliph and many others. **8.5″ X 11″, 94 ppgs, Retail Price: $10.99**

Wyoming Mining Books

Mining in the Laramie Basin of Wyoming - Unavailable since 1909, this publication was originally compiled by the United States Department of Interior. Also included are insights into the mineralization and other characteristics of this important mining region, especially in regards to coal, limestone, gypsum, bentonite clay, cement, sand, clay and copper. **8.5″ X 11″, 104 ppgs, Retail Price: $11.99**

More Mining Books

Prospecting and Developing A Small Mine - Topics covered include the classification of varying ores, how to take a proper ore sample, the proper reduction of ore samples, alluvial sampling, how to understand geology as it is applied to prospecting and mining, prospecting procedures, methods of ore treatment, the application of drilling and blasting in a small mine and other topics that the small scale miner will find of benefit. **8.5" X 11", 112 ppgs, Retail Price: $11.99**

Timbering For Small Underground Mines - Topics covered include the selection of caps and posts, the treatment of mine timbers, how to install mine timbers, repairing damaged timbers, use of drift supports, headboards, squeeze sets, ore chute construction, mine cribbing, square set timbering methods, the use of steel and concrete sets and other topics that the small underground miner will find of benefit. This volume also includes twenty eight illustrations depicting the proper construction of mine timbering and support systems that greatly enhance the practical usability of the information contained in this small book. **8.5" X 11", 88 ppgs. Retail Price: $10.99**

Timbering and Mining - A classic mining publication on Hard Rock Mining by W.H. Storms. Unavailable since 1909, this rare publication provides an in depth look at American methods of underground mine timbering and mining methods. Topics include the selection and preservation of mine timbers, drifting and drift sets, driving in running ground, structural steel in mine workings, timbering drifts in gravel mines, timbering methods for driving shafts, positioning drill holes in shafts, timbering stations at shafts, drainage, mining large ore bodies by means of open cuts or by the "Glory Hole" system, stoping out ore in flat or low lying veins, use of the "Caving System", stoping in swelling ground, how to stope out large ore bodies, Square Set timbering on the Comstock and its modifications by California miners, the construction of ore chutes, stoping ore bodies by use of the "Block System", how to work dangerous ground, information on the "Delprat System" of stoping without mine timbers, construction and use of headframes and much more. This volume provides a reference into not only practical methods of mining and timbering that may be employed in narrow vein mining by small miners today, but also rare insights into how mines were being worked at the turn of the 19th Century. **8.5" X 11", 288 ppgs. Retail Price: $24.99**

A Study of Ore Deposits For The Practical Miner - Mining historian Kerby Jackson introduces us to a classic mining publication on ore deposits by J.P. Wallace. First published in 1908, it has been unavailable for over a century. Included are important insights into the properties of minerals and their identification, on the occurrence and origin of gold, on gold alloys, insights into gold bearing sulfides such as pyrites and arsenopyrites, on gold bearing vanadium, gold and silver tellurides, lead and mercury tellurides, on silver ores, platinum and iridium, mercury ores, copper ores, lead ores, zinc ores, iron ores, chromium ores, manganese ores, nickel ores, tin ores, tungsten ores and others. Also included are facts regarding rock forming minerals, their composition and occurrences, on igneous, sedimentary, metamorphic and intrusive rocks, as well as how they are geologically disturbed by dikes, flows and faults, as well as the effects of these geologic actions and why they are important to the miner. Written specifically with the common miner and prospector in mind, the book will help to unlock the earth's hidden wealth for you and is written in a simple and concise language that anyone can understand. **8.5" X 11", 366 ppgs. Retail Price: $24.99**

Mine Drainage - Unavailable since 1896, this rare publication provides an in depth look at American methods of underground mine drainage and mining pump systems. This volume provides a reference into not only practical methods of mining drainage that may be employed in narrow vein mining by small miners today, but also rare insights into how mines were being worked at the turn of the 19th Century. **8.5" X 11", 218 ppgs. Retail Price: $24.99**

Fire Assaying Gold, Silver and Lead Ores - Unavailable since 1907, this important publication was originally published by the Mining and Scientific Press and was designed to introduce miners and prospectors of gold, silver and lead to the art of fire assaying. Topics include the fire assaying of ores and products containing gold, silver and lead; the sampling and preparation of ore for an assay; care of the assay office, assay furnaces; crucibles and scorifiers; assay balances; metallic ores; scorification assays; cupelling; parting' crucible assays, the roasting of ores and more. This classic provides a time honored method of assaying put forward in a clear, concise and easy to understand language that will make it a benefit to even beginners. **8.5" X 11", 96 ppgs. Retail Price: $11.99**

Methods of Mine Timbering - Originally published in 1896, this important publication on mining engineering has not been available for nearly a century. Included are rare insights into historical methods of timbering structural support that were used in underground metal mines during the California that still have a practical application for the small scale hardrock miner of today. **8.5" X 11", 94 ppgs. Retail Price: $10.99**

www.ingramcontent.com/pod-product-compliance
Lightning Source LLC
Chambersburg PA
CBHW080250180526
45167CB00006B/2480